VOLUME 7 IN THE SERIES
Our Sustainable Future

Series Editors

Lorna M. Butler
Washington State University

Cornelia Flora
*Virginia Polytechnic Institute and
State University*

Charles A. Francis
University of Nebraska–Lincoln

William Lockeretz
Tufts University

Paul Olson
University of Nebraska–Lincoln

Marty Strange
Center for Rural Affairs

Huey D. Johnson

Green Plans

Greenprint for
Sustainability

with a new afterword
by the author

University of Nebraska
Lincoln and London

Manufactured in the United States of America.
⊖ The paper
in this book meets the minimum requirements of
American National
Standard for Information Sciences – Permanence
of Paper
for Printed Library Materials, ANSI Z39.48-1984.
First Bison Books printing: 1997
Most recent printing indicated by the last digit below:
10 9   8   7   6   5   4   3   2   1
Library of Congress
Cataloging-in-Publication Data. Johnson, Huey D.
Green plans:
greenprint for sustainability / Huey D. Johnson;
with a new afterword by the author.
p.   cm. –
(Our sustainable future; v. 7)
Includes bibliographical
references and index.
cl: ISBN 0-8032-7596-X (pa.: alk. paper) 1. Environmental
policy. 2. Sustainable development. 3. Green movement.
I. Title.   II. Series.
GE195.J64   1997   363.7 – dc20
96-35373   CIP

Contents

Illustrations

Figures

Maps

Photographs

Foreword

By David R. Brower

Too often called the grandfather of the environmental movement, I consider myself a grandson. The movement has been in existence for a long time. It was going strong in 1864, when Yosemite was set aside as the first park for the nation. We've learned a great deal over the years and have brought about many important changes. Environmental restoration efforts are succeeding in many parts of the world, but they just aren't going fast enough to catch up with the trashing.

Look at the vital signs on the state of the earth: all indications show it is going downhill. In my lifetime, eighty-two years, the population of the earth has tripled, and that of California has gone up by a factor of twelve. In my lifetime we have used four times as many resources as in all previous history. In the Great Valley of California we had six thousand miles of salmon streams, and now we are down to two hundred. In the great Sierra Nevada we had four Yosemites, now down to three.

In the last twenty years, one-seventh of the earth's cropland has been turned to desert, or worse. Human population has doubled. People have destroyed enough forest to cover the United States from the Mississippi River to the Atlantic seaboard. They have driven nearly a million species of plants and animals to extinction, and spent more natural capital than in all previous history.

Tom Hayden has said that all he has been able to do in his career is slow the rate at which things get worse. I have come to realize that that was true of me, true of the environmental movement, true of too few corporations, and too many universities. One of my favorite quotes comes from Alwyn Rhys, who said long ago in Wales, "When you've reached the brink of an abyss, the only progressive move you can make is to step backward." I told that story twenty years ago in southern New Zealand, and someone said, "Well, restate it. Just say, when you've reached the edge of an abyss, the only progressive move you can make is to turn around and step forward."

There is support out there for such a turnaround, a U-turn away from the brink. In 1993, the Union of Concerned Scientists published a statement signed by two thousand scientists, a good hundred of them Nobel laureates. It says, "No more than one or a few decades remain before the chance to avert the threats we now confront will be lost, and the prospects for humanity immeasurably diminished. A new ethic is required, a new attitude toward discharging our responsibility, for caring for ourselves and for the earth. This ethic must motivate a great movement, convincing reluctant leaders and reluctant peoples themselves to effect needed change." That is a statement from two thousand scientists – many more than the number of scientists supporting Rush Limbaugh's Pollyannish position.

What we need to do is completely rethink our situation, and remember that conventional wisdom has brought us to where we are. It is time to give the only living planet we know some dynamic CPR – conservation, preservation, and restoration. Conserving resources by rational use, preserving those we cannot replace, and restoring natural and human systems. Regenerated natural capital can secure a sustainable economy and society.

This whole idea of conservation, preservation, and restoration is not altogether new, and there have been many organizations working on it. We do not have to invent it; there have been many successes in restoration, but not nearly enough. We need a mobilization for it. Back in Franklin Roosevelt's days, to get out of the great economic depression, we created the Reconstruction Finance Corporation, the National Recovery Administration, the Public Works Administration, the Works Progress Administration, and many others. Now we've got to get out of the environmental depression we are in, because we've been spending natural capital as if there were no limit to it.

Huey Johnson is an artesian well in a world where that kind of well is increasingly hard to find. His mind is a watershed – to mix metaphors – that has been cared for better than most. It is inevitable that such a constant flow of fresh ideas would nurture the most satisfying color of all – green.

I should warn readers of an ever-present danger. Huey Johnson's ideas are quite contagious. Allow me to cite myself as an example. Having been in the conservation business for more than half a century, I was unbelievably slow in discerning the importance of restoring natural and human systems. Huey doesn't particularly like the word restoration and rarely uses it, but that is what he is all about.

Consider the program of restoration he initiated in 1977, when he was secretary of resources for California governor Jerry Brown. He describes it in this book, but not at great enough length, considering its importance to the worldwide green plans he is now generating. Let me describe the California origin here, in due brevity, as I see it.

Huey's "Investing for Prosperity" program would be the precursor of his Green Century Project, the most dramatic of the ideas being developed by the New Renaissance Center he founded (now renamed the Resource Renewal Institute), spurred on by what he had learned from The Nature Conservancy and from his innovative work in Trust for Public Land, which he also founded.

For the State of California Huey advocated a hundred-year plan to improve the state's productivity and environmental quality. Finding his friends laughing at the idea, he went to his ostensible enemies – California's corporations.

As he tells it in *Earth Island Journal* (Spring 1987), the first stop was IBM, where one of the highest-ranking officers agreed to help. The work of IBM was to take detail and drudgery out of people's lives, and IBM was certainly interested in enhancing long-term economic and environmental quality. Huey asked, "Can we borrow your lobbyist?" And IBM said, "Help yourself."

Huey then went to Southern Pacific, Bank of America, and other corporations with whom he had been brawling from time to time. They were enthusiastic. Next he went to labor unions, who loved the idea because it would create jobs and enhance the quality of people's lives. The League of Women Voters made Investing for Prosperity their own statewide project.

"Suddenly everybody we talked to wanted to get involved," Huey said. "Then we signed on the environmentalists."

He asked for ideas about what to include in the one-hundred-year plan, and found that people liked to be asked. Some suggestions, he said, were excellent, some off-the-wall, some unpredictable. He put thirty-five projects together and won an appropriation of $125 million a year from state-owned offshore oil revenues.

His Green Century formula included similar hundred-year plans for every part of the system – conservation and improvement of soil properties, coastline, parks and recreation areas, wildlands, and energy projects – all based on organizing a broad-based constituency for long-term plans for upgrading environmental quality. These were developed by skilled, nonpartisan individuals, and funded with government revenue from publicly owned nonrenewable resources, and also from such renewable resources as forests.

Investing for Prosperity was also the precursor of the green plans Huey Johnson talks about in this book; they are based on the same principles and concepts. The Netherlands has a twenty-five-year program, and every four years the government, the corporations – the whole lot – are required to account publicly for how they are getting along. They are restoring land that is now in agriculture to the marshland it once was – about 10 percent of the agricultural land. They have a $500 deposit on cars, and that means whoever gets the old relic and takes it apart and recycles it has $500 to play with. It is an important plan, and we should all learn more about it. Canada is supposedly getting such a plan going, and New Zealand has a very important one, where the boundaries for management are ecological, not political, boundaries. These are opportunities for restoration, and it is important to look closely at them.

In Huey's view, if you link up all the people interested in air, agriculture, water, energy, forestry, and wildlife, for example, you have a powerful political base that's determined to deliver a positive future. "When you get them all together," Huey concluded, "it's just like sending an avalanche through a tea party."

I have nothing against tea parties, but I like artesian wells better than avalanches. And what flows from the Huey Johnson well ought to be carried, by all routes, surmountable or insurmountable, to places that have lost their green.

Robinson Jeffers had it right when he said:

> It is only a little planet
> but how beautiful it is.

In the few decades since he wrote those lines, the *is* has been turning to *was* at an increasingly rapid rate. So this is the question about the only living biosphere we know: is it to be or not to be?

Herein, Huey Johnson shows us how to put, and keep, the "be" in our bonnet.

Acknowledgments

First and foremost, my thanks and deep appreciation to writer Eileen Ecklund. Eileen worked with me practically from the beginning of this book, critiquing my arguments and editing my prose. This book would not have been possible without her exceptional skills, her passion for the subject, and her good humor. However, I take full responsibility for the result.

There are many others I would like to thank who generously contributed their time and thoughts to the contents of this book. Among them are Peggy Lauer, Michael Painter, Tyler Johnson, Phillip A. Greenberg, Erika Bley, Jake Kosek, A. Michael Signer, Ann Kelly, Paul Hofseth, Lindsay Gow, Roger Blakeley, Tom Fookes, Craig Lawson, Hans van Zijst, Eric Brandsma, Robert Currie, Alfred Heller, Sylvia McLaughlin, John Skov, Jessica Presson, and Jason Morrison. Their efforts and insights have been invaluable.

My thanks and appreciation also go to the green planners and other government officials and their staffs, the environmentalists, and the businesspeople of the Netherlands, New Zealand, Canada, the United States, and all the other nations mentioned herein, for taking the time from their busy lives to explain their countries' policies to me, and for having the vision and courage to carry out their nations' green plans.

Last but most certainly not least, I would like to thank the philanthropic foundations and individuals who have supported the idea of green plans and who helped make this book possible: the Wallace Alexander Gerbode Foundation for its loan, the Clarence E. Heller Charitable Foundation, Fred and Annette Gellert of the Gellert Foundation, the Lyddon family and the Seven Springs Foundation, the Marin Community Foundation, the Columbia

Foundation, Max Thelen Jr. and the S. H. Cowell Foundation, the Frank Weeden Foundation, the Wheeler Foundation, Henry Corning, Marion Weber, Dan Hewitt, Gil Ordway, Doug Carlston, Bruce Katz, Anne Pattee, Nancy Kittle, Jim Compton, Patagonia, Inc., Malinda and Yvon Chouinard, Christian Erdman, and Helen Dreyfus.

Introduction

This is a book about solving the environmental problem in a way that works for the world, the nation, businesses, labor, environmentalists, future generations – for everyone. It is about green plans, or national environmental strategies, which I believe are the path that nations and regions must take to respond to environmental decline. Green plans approach the problem in a way we have never approached it before, this time in a serious effort to *solve* it. They are about believing that we can put the problem behind us – as we put polio behind us – and in the process, find that we all benefit when we work together.

Green plans are also about rescuing the concept of planning from the scrap heap of history. Planning seemed so good and so important back in the 1950s and 1960s, but then it simply became an excuse for not making a political decision, and the enthusiasm for it faded abruptly. I realize now that, in terms of the environment, the difficulty with planning was that we were not looking at the problem on a large enough scale. We did not have a structure that was comprehensive enough to do what had to be done: approach the problem with the intention of solving it.

Green plans show what planning can and should be. They are comprehensive, integrated, and large-scale – three characteristics that are key to solving environmental problems, whether on the local, regional, or national level. I have studied endless alternatives and read what now seems to be thousands of proposals and philosophical discourses, all discussing topics related to this central dilemma of declining environmental quality. But in all those documents, I do not remember one that looked at the issue in terms of its actual scale and complexity, taking into consideration the entire set of relationships among air, soil, water, plants, animals, and people.

We have been putting bandaids on broken legs. Our response to environmental decline has been to approach it one part at a time: air quality this year, toxics last year, forestry for a few weeks next summer. A watchmaker cannot fix a watch by tinkering with only one part of it, and we cannot solve water pollution without working on acid rain, air pollution, and sustainable energy policies at the same time.

Our world consists of living ecological chains, each mutually dependent upon and influencing each other. If we are to salvage our planet's future, we must also link our environmental problems, and re-create the whole picture. Rene Dubos, the renowned biologist and philosopher, said that ecology may be more complex than the human mind can understand. That may be true, but now, with the computer and the growing body of knowledge we have to work with, we can at least begin to think more ecologically than we have been able to in the past.

Because our old approaches fell far short of what was necessary, the public has lost a lot of faith in its leaders. Although there have been some significant accomplishments, the pressures on the environment keep growing and the overall situation is deteriorating. The green plan idea, which is just beginning to emerge in such nations as the Netherlands, Canada, and New Zealand, is a new source of hope.

Green plans are quite unlike anything we have done before, both in scale and in importance. But despite the massive dimension of the project, green plans are not a theory; they are actually happening. They are experiments that are being conducted in the real world, with real, measurable results.

The individual pieces that go into making up any country's green plan are not revolutionary; most are not even new. What is so radical about what these countries are doing is the scope of their vision, the fact that they have pulled together all the related pieces into one package. They have accepted that natural resources are a complex, interrelated system and that any real environmental plan must be comprehensive enough to embrace that complexity. They have made environmental recovery their top priority, and have set into motion large-scale efforts guided by the government and involving all segments of society.

In addition to being comprehensive, integrated, and large-scale, green plans are also based on the critically important premise that our social and economic well-being depend on a healthy environment, and that we must manage our natural and physical environment in a sustainable fashion if we want to continue to meet our own needs and to allow future generations to meet theirs.

Green plans are government funded and implemented, but develop strategies for working with the private sector to reduce negative environmental impacts. A green plan needs the support of a broad, diverse public constituency. It must be based on long-term planning, and be flexible enough to change as new information becomes available.

A green plan is developed by government with the input of key players in the region it covers – not mandated from the top down – and is designed to deal with problems appropriately at each level of scale. For example, both the Netherlands and New Zealand delegate the responsibility for managing local environmental problems – in fact, for implementing much of their respective green plans – to local authorities. But both nations also set standards at the regional and national levels, because they realize that many environmental issues cross the boundaries of city and region. The same will be true in the United States. While cities and states will implement their own green plans, federal standards and agreements will be necessary to deal with issues that cross boundaries. Ultimately, some environmental issues will require international agreements and standards.

Because green plans are comprehensive, no one popular issue, such as wilderness preservation, rainforests, or toxics, dominates the overall strategy. Each issue is an important part of the whole, and each gains more clout *because* it is part of a larger plan. All the different issues and their constituencies are forged into a coalition that has as its goal the resolution of all problems. As a result, more people can relate to the many interests, and the plan gains popular and political support.

The power of green plans is in their scope. People take hope from such a large-scale effort because they can see that their goverment is serious in its commitment. The threat is thus turned into an asset, bringing people together and helping them to put conflict behind them. People are willing to make small sacrifices when they can see how big the payoff will be – a livable future for their grandchildren – and when they understand that their entire society is working toward the same end.

Such a dream, of course, requires focused, disciplined, hard work, including the political work of selling the idea and involving people as a force. Only the people can cause government to make environmental recovery its first priority. This shift in priorities is essential: We have the need for environmental recovery, and we have the technical knowledge and the resources to accomplish it; but in order to do so we must make it a priority item on our list, as we have in the past for bombs and rockets and space vehicles.

Time is of the essence. Each day we delay in moving toward a plan for comprehensive environmental recovery is another day in which our natural resources are degraded even further. But we do not have to reinvent the wheel; we can and should take advantage of the years of painstaking work that other countries have already devoted to developing green plans. We can learn from their mistakes and their successes, adapting what is best from each plan to our own needs.

The purpose of this book is to define green plans as they exist, to present the case for them to the public, and to serve as a central source of information for politicians, planners, students, activists, and anyone else interested in the process of green planning. If it seems that the focus of this book is primarily on the developed nations, it is because they have created the biggest problems of environmental degradation, and it is they who have the most work to do in the coming years.

To this project I bring the benefit of my own twenty-five years of experience as an environmentalist and planner. The information in this book also reflects the work of the Resource Renewal Institute (RRI), which I founded in 1983 to study and promote green plans. Recently, RRI established the Global Green Plan Center to serve as an information clearinghouse and international training center for the new policy field of green planning. The Global Green Plan Center will use the latest in communications technology, including the Internet, to link government officials, planners, and environmental and community orgainzations worldwide. The Center's online information can be accessed through the World Wide Web (http://www.rri.org); its e-mail address is info@rri.org.

The destruction of the global environment is an enormous threat to our security as a nation and as a species. Our response to it must be as great as if we were preparing for a third world war – nothing less will do. But the effort we are engaging in is the reverse of a war, because people and nations will work together toward recovery, not destruction. It all starts with us as individuals, taking the first step and then carrying it through.

Some will respond with a long list of reasons why we cannot possibly solve the problem of environmental decline. While it is possible to refute those arguments, the reaction is nonetheless indicative of the psychological barriers many have erected. It often seems as if we have accepted the degradation of our environment as a hopeless situation. It is my belief that the example of the pioneering green plan nations can light a spark in those without hope; it is in that spirit that I offer this book.

PART ONE

Defining the Problem and Its Solution

1

A Commitment to Change

For many years we have known that some human activities damage the environment, and ultimately ourselves. When the problems caused by these activities have become too extreme to ignore, as when the air has become too dirty to breathe or the water too polluted to drink, we have attempted to fix them.

But repairing the damage has become more and more difficult as our population has increased and the needs of these new millions have multiplied the pressures on the environment. The problems are no longer local but regional and continental; now we are even affecting global systems like the ozone layer. The dying oceans are a clear indication of the gravity of the problem we are confronting.

The Mediterranean fisheries, once so prolific, by and large no longer exist. The Pacific coast of North America was once home to a thriving sardine fishery employing twenty-five thousand people in California alone; today there is no commercial sardine fishing at all. Only the novels of John Steinbeck remain to remind us of humanity's impact on a once-great resource.

We are also seeing the effects of overexploitation of land-based resources. Countries that use intensive crop irrigation methods are suffering a loss in soil productivity because of the build-up of salt residues these practices cause. We would do well to remember that this was also one of the reasons for the decline of the Roman Empire.

We have ignored scientific warnings about pollution and overexploitation, and as a result have pushed some resources beyond their capacity to recover. In some areas, there is no longer food to eat and no work available in formerly productive, resource-based industries.

What are we to do about these problems? In the past we have always looked at one issue at a time, passing air pollution legislation one year, devoting some funding to endangered species the next. But this piecemeal approach is not working. Ultimately, our goal must be to fashion a society that is able to function within the limits set by nature. We must manage in a way that will lead to a sustainable society, not just tinker with parts of the problem, as we have been doing. We must adopt large-scale, comprehensive, integrated plans that are designed to *solve the problem* in its entirety.

To advocate this new approach is not to belittle the extraordinarily valuable efforts made by so many people for so many years in fields as diverse as science, agriculture, technology, and finance. That we are now able to make such a leap is due in large part to their work. It is also due to the struggle of those who have sought to increase public awareness of environmental issues. Because of them, many of the world's people now accept the seriousness of the problems and are ready to undertake the major efforts necessary to halt and reverse the damage.

Because it is so new, the idea of solving the entire problem of environmental decline may sound impossible to many people. How can we do that if we cannot even make much progress on the smaller, individual issues? But that is precisely the power of a big-picture approach: by tackling the larger problem, you resolve a host of smaller ones.

Comprehensive plans are the only way to solve larger-scale problems because they look beyond the individual issues to the problems created by the relationships *between* those issues. The reality of ecology is that we cannot solve the individual problems unless we include the relationships of each to the other. We have the tools we need to move forward with this kind of plan; what we need is the will to do it.

It seems that no one believes a project is doable until there are working models of success. The standard response from the "experts" is always: "If that kind of approach worked, somebody would already be doing it." In the case of green plans, we can now answer that somebody is. A few comfortable, developed nations – ones that have traditionally been in the forefront of progressive social change – have begun to lead a global movement toward environmental recovery.

New Zealand was the first nation to give women the right to vote. The Netherlands was one of the first nations to have child labor laws. These two nations and Canada are now out in front of the rest of the world in solving the environmental problem. What they are doing is truly revolutionary: they have taken the position that a solution is not only possible, but essential, if we are to leave anything at all for future generations. Each of these countries has adopted its own comprehensive environmental policy, or green plan, a practical strategy designed to translate the concept of sustainability into action on the national and local levels. I believe the rest of the world will soon follow, with each nation developing its own innovations.

There is a difference between musing about something and actually doing it, between thinking about what is possible and making those possibilities real. Throughout history there have been instances of revolutionary ideas that were pondered and discussed for years, building and building until some person or group finally brought them to life. The idea of human flight is one such example. From the time of the ancient Greeks up until the early twentieth century, humans speculated and dreamed about the possibility of taking to the skies, but very few believed it would ever happen. When it finally did, it changed human society.

Green plans are the beginning of another such leap. The countries that are implementing them are the first in history to attempt to recover environmental quality nationwide. They have made environmental sustainability a key issue of national purpose. Like the Wright brothers, they are pushing humanity over the threshold of a new era.

We can spend generations pondering a concept – discussing it, designing it, refining it – but until it is actually put into action, until it becomes real, it is no better than a hundred other ideas. The Netherlands, New Zealand, and Canada have let go of the safety line and moved from theory to practice, with all the risks inherent in that. Because they have taken the initiative, the rest of the world will be able to benefit from their experience, observing their successes and setbacks, learning what works and what does not.

At this point, it would make no sense for any city, state, or nation to start from scratch in creating its own plan, because the level of thought and the degree of commitment that has gone into the plans of the Netherlands, Canada, and New Zealand is remarkable. Starting from scratch would cost many years and many millions of dollars, and it is unnecessary. We are not going to invent a better wheel. What these countries have done will be the models for others to follow and build upon. Each country will want to adapt its plan to its own circumstances, but the basic green plan design will be like that of the two-wheel bicycle: While individual bikes vary widely, the design principles followed by most have changed very little in the past century.

Each of these three countries' green plans will be described in detail in separate chapters, and examples from them used to illustrate points throughout. It is important to keep in mind, however, that each country's plan is at a different stage of development. Because the Netherlands' plan was the first to be passed and is the most completely implemented so far, there is much more information available about it, and it thus plays a larger role in this book. New Zealand's plan is currently being implemented, but many of the day-to-day aspects have yet to be worked out. The Canadian plan has been slowed, but not stopped, by economic conditions and the recent political changes there.

Each of the green plans designed by the pioneering countries is unique, taking into account that nation's distinctive characteristics and problems. But just as there are certain basic elements to a clock – the mainspring, the hands, the gears – there are certain elements basic to green plans. These are comprehensiveness, integration, and a large-scale commitment by government. Green plans share other elements as well, primarily because they are examples of systems thinking – looking at whole systems rather than at discrete parts (part 3 covers some of these other elements in depth). But, taken together, the three "mainspring" elements constitute the definition of green plans as they have evolved to date; these are discussed in detail below.

Comprehensiveness and Integration

When we talk about the environment, we are not talking about just trees, or water, or air, but of all those things and more, interrelated in a very complex system. The ways in which we as humans interact with that system are equally complex: extracting resources, irrigating farmland, harvesting trees, burying our waste, creating energy. We cannot hope to remedy the effects we have had on the planet unless we develop policies that use this complexity as their starting point.

Our approaches to resource management and the environment in general have been fragmented. Forestry is an example: Most nations have traditionally been concerned solely with the economics of forestry, whether in India one thousand years ago or the United States over the last hundred years. From the point of view of economics, forestry means cutting trees, creating finished wood products, building homes and businesses, and carrying on trade. As forest resources rapidly depleted, some concern was shifted to the idea of harvesting trees sustainably, but even this approach generally failed to take anything but trees into consideration. Recently, however, some regions have realized that their fisheries are dying out, in part because silt from eroding clear-cut slopes has affected spawning streams. This discovery has led to the realization that forestry and fisheries are linked.

We have also, in recent years, discovered links between forests and air pollution. The trees in some of the world's great forests, including Germany's Black Forest, are dying because of air pollution. When the trees die, so do the songbirds and all the other life forms that depend on them, from microbes to elk. This has made us understand that if we want to have forests, we have to be concerned about the ways in which we are diminishing the quality of the air, from the toxics spewed from smokestacks to the exhaust belching from tailpipes. A third of the forests of Europe are suffering from the effects of air pollution, and it is increasingly affecting those of Canada and the United States.

Yet government efforts are still rarely organized to manage resources as a system; instead, they are typically fragmented among dozens of different agencies, each dealing with a single issue. In order to survive they engage in turf wars, fail to coordinate their policies, and fight over scraps of funding. Looking at a dozen agencies of government dealing with the environment – wildlife, parks, forestry, soil, water quality, and so on – reveals that they are rarely managed by one administrator, nor are they operated as a cohesive unit, functioning together the way a clock works.

Politics and power decide how various environmental issues are ranked. In most state or federal agencies in the United States today, agriculture, oil, and water are probably ranked at the top. Soil, energy, wildlife, parks, and recreation get little attention or funding. The powerful agencies guard their privileged positions jealously, while the less powerful are left with the crumbs. They are fighting each other needlessly, and often not managing their affairs very well.

When we leave behind an issue that is underfunded, like soil, we undermine all our efforts over the long term. We are wasting our money whenever we deal with forests and air quality and don't deal with the other issues. Often we are just pushing problems around from one realm to another, cleaning the water only to bury or incinerate the contaminants that have been removed, thereby polluting the soil or the air.

If we were to compare our attempts to understand and improve environmental quality to our concern about the health of our own individual bodies, we might say that to date we have been looking at one foot and little else. We know, of course, that there are other problems, of bones and blood and diet and much more. When we look beyond the primitive approach we have taken to the environment, we see living concepts that are part of the earth, all interconnected and interrelated.

The government of the Netherlands has put its finger on the problem very concisely: "The difficulty with this fragmented approach is that it addresses a succession of new issues without necessarily resolving the previous one, thereby creating the impression that it no longer matters. Attention focuses on one subject, overshadowing others which are no less important. This approach also fails to treat the environment as a single system, which makes it virtually impossible to show people how their behaviour affects the environment."[1] By contrast, the Netherlands' comprehensive program aims to make environmentally friendly behavior second nature to its people.

To achieve environmental recovery, we need to accept complexity and operate in a systems environment, making systems decisions. Green plans are able to address this complexity, first of all because they are *comprehensive,* embracing all environmental and resource issues, across media and across geographical boundaries. Second, green plans are *integrated* throughout human society as it relates to the environment, from industry and government to social groups and individuals. Green plans look at the interconnections and relationships between different environmental issues and between natural and human systems, and create similar links between those responsible for creating and implementing environmental policies.

Comprehensive, integrated approaches to environmental planning bring cohesiveness to government efforts, encouraging coordination and cooperation. They also make government and the management of environmental quality understandable and sensible to the public.

The green plans of the Netherlands, New Zealand, and Canada are comprehensive, ecosystem-based initiatives designed to save the forest, not just the trees. Instead of passing laws that attack each problem one by one in isolation, these countries have created approaches that cut across traditional lines in ways that make sense for their resources, population, and industry. And within

14

government itself, they have pulled together all the major ministries and agencies into one coordinated effort to achieve environmental quality.

Implementing such plans is obviously a challenge for each country. However, their comprehensiveness has in some ways made implementation easier, allowing all three nations to move away from the layering of regulatory and legal approaches that had developed over the years. They have replaced this old system with a refined, more efficient, broader strategy that gives businesses and individuals greater latitude to meet and maintain environmental quality goals. This means far less frustration, particularly for businesses that have tried to cooperate in the past, only to become mired in overlapping or outdated regulations.

An important element in this sort of comprehensive planning is to bring together all interested groups, including environmentalists, industry, and citizens, to carefully review existing laws and regulations and develop a new approach. The objectives must be clearly established and the limits clearly defined; once that is done, it is no longer necessary to pile regulation on top of regulation.

It is interesting to compare what is happening in these three countries to the policies of California, a state that has a reputation for being very modern and efficient, and which has adopted strong, farsighted policies on a number of environmental issues. For example, California led the energy revolution, and as a consequence the state has a steady stream of visitors from all over the world who come to study its energy efficiency model. It also has strict air quality standards. But because California has not made the leap to the broader, comprehensive approach, its policies in other areas are severely lacking. The state's water policy is backward and poorly managed, and soil policies scarcely exist at this point. The same is true for rapid transit and growth policies. Until California links its policy and regulatory programs together, it will not come close to providing a livable future for its citizens.

The same flaw is also evident in nations such as Germany and Japan, which are very efficient with some issues but fall short in others. Until a nation or city or state embraces the comprehensive approach, anything it does will be less efficient and more expensive.

Environmental strategies for the future will have to be comprehensive in order to cope with the complexity of the environment and of the problems we face. And comprehensive plans will *work*, because they have the power of

15

mass behind them. An environmental idea has a better chance of success if it is part of a larger whole; single issues are far easier to block or defeat.

A Large-Scale Commitment by Government

The other critical element of green plans is their scale: the size and scale of the project must match that of the problem. In the past, we have not really comprehended just how big and complex environmental problems are, so we have not responded appropriately.

Scale is one reason that green plans place responsibility for environmental planning at the national or state level; only government can manage something of this size efficiently and effectively. For instance, just the amount of money that is required – billions in the case of existing national plans – is far beyond what an organization or institution would be able to provide.

There are other green plan functions that only government can do, such as managing the taxes and environmental quality regulations, and enforcing the laws. Only government has the resources and the scope required to handle a project this immense. If it is efficiently run, the government has a tremendous advantage when it comes to the delivery of certain services.

In the United States in the last few decades, it has become fashionable to think that government is useless, and that only private enterprises can handle problems efficiently. But it is a mistake to view the two as necessarily opposed. Privatization can be an important part of the way a government functions, but it should not be seen as a panacea that absolves the government of responsibility. For example, it may be most efficient to have a private firm collect a city's trash, but it is still the government's responsibility to make sure that the trash *is* picked up, and disposed of properly. So while green plans may well include roles for private enterprises, they will require, first and foremost, government leadership and commitment to the goals to be achieved.

We have always underestimated the environmental problem, but we come closer to understanding its size and scale when we see the size and scale of the response from a nation like the Netherlands. The Netherlands' green plan is an example of technical excellence, a multifaceted and detailed approach. To achieve its goal of recovering environmental quality in twenty-five years, the government has enlisted hundreds of people, including some of the country's best minds.

What is happening now in the Netherlands reminds me of the preparation for the Normandy invasion during World War II. I remember well the pictures of thousands of ships sitting offshore – the largest armada in history, all coordinated, all waiting to act in concert. It is that kind of human endeavor that a nation undertakes with these green plans: a massive commitment to a purpose.

Looking back in U.S. history for a comparison, perhaps the best is with the soil conservation effort intended to stop wind erosion of the Great Plains at the end of the Depression. For a very short time, soil was understood as the key to civilization's survival, and great passion went into the effort to save the Plains from devastation. But the effort was not maintained over the years, and in any case fell short of what was needed. It cannot be compared to the complexity and size of what is happening in the Netherlands today.

The scale of our *funding* commitment has to match the size of the problem, too. We have a habit of putting a symbolic amount toward an issue, rather than enough to really have an effect. In trying to determine whether an institution is committed to a particular policy, one of the best questions to ask is what portion of the total amount it spends is devoted to implementing that policy. A government or corporation can say that it has the best health program in the world, but if it has yet to commit any money, then all it has is a piece of paper, not a program.

Using this measure, the Canadian example is quite remarkable. At a time when its economy was battered by recession, the Canadian government pledged to spend $2.2 billion over the next six years.[2] The Dutch estimate they will spend $9.5 billion on the environment in 1994; this is almost twice what they spent per year before the green plan was adopted.[3] These amounts, large as they are, are even more remarkable for coming from rather small treasuries. The immensity of the commitment on the part of these countries gives their plans an integrity and seriousness of purpose that smaller efforts simply cannot match.

In the past, we have rarely been able to think far enough ahead to adequately fund a project. As a result, we often end up wasting money and falling short of our goals. One example is the case of Redwood National Park in California. When the government first started buying property for the park, a few hundred acres at a time, the land was available for $500 million. But it took about ten years to purchase the amount desired, and over that time the cost increased to more than $1 billion.[4] The park cost at least twice as much as it would have if it had been properly financed from the beginning, and the land bought all at once.

The scale of the problem was creating Redwood National Park, with its tens of thousands of acres, and the government approached it a tree at a time. Had it not been for the efforts of the nonprofit Save the Redwoods League, which had been buying and preserving land for decades prior to the creation of the park, it would have been even more costly. The partnership between the government and the nonprofit worked well in the end, but much money could have been saved had the government had the foresight to commit enough funding to begin with.

If a national environmental recovery strategy is to be successful, it must incorporate and build upon the three main principles of comprehensiveness, integration, and a large-scale commitment by government. The precise methods used to implement these principles will be different for each nation, and will probably change over time, but a program that fails to include any one of them will not be a true green plan, and its chances for success will be fewer.

• • •

The examples provided by the Netherlands, New Zealand, and Canada are especially important now that the world is shifting into a new phase in environmental planning. The adoption of Agenda 21 by more than 170 nations attending the UN's Earth Summit in Rio in 1992 indicates widespread recognition that our old ways of responding to environmental problems are no longer sufficient. By adopting Agenda 21, those nations agreed to follow a comprehensive, integrated, green plan approach to managing their environmental affairs. The pioneer green plans will be extremely useful models for the nations now starting down this track.

2

Sustainability from Theory to Practice

The words that launched many of the ideas important to green planning were "sustainable development." Although the concept has been a subject of academic study for decades, it really burst onto the international political scene in 1987 with *Our Common Future*, a highly influential report of the UN's World Commission on Environment and Development. Called the Brundtland Report after the commission's chair, Norwegian Prime Minister Gro Harlem Brundtland, *Our Common Future* emphasized the links between problems of growth, economics, technology, and the environment. As a solution it proposed sustainable development, which it defined as "development that meets the needs of the present without compromising the ability of future generations to meet their own needs."[1]

Ever since the Brundtland Report was published, sustainable development has been a widely discussed and debated concept. The report itself and the debates it inspired gave shape and impetus to the green plan idea in the countries that are now adopting it. All green plans have the ultimate goal of achieving sustainable development, but though the term itself serves an important philosophical and social purpose, it is limited in a number of ways.

Its primary limitation is that it is so vague as to allow a multitude of definitions. To some extent, "sustainable development" has become a buzzword, often put forth as the solution to humanity's problems but rarely accompanied by any explanation of how it is to be accomplished. It has come to be used as one would use words like philosophy or religion – yes, the world needs philosophy; yes, the world needs sustainable development – but is meaningless in terms of defining necessary actions.

Regardless of the debate over terminology and application, sustainable development has attained a remarkable popularity and significance, linked as it is to growing concerns about the global impacts of environmental decline. To date, none of the attempts to refine the concept has had a more dramatic effect on world thinking than Brundtland's original definition, that of meeting current needs without compromising the ability of future generations to meet theirs. And although it has not solved, and cannot solve, all our problems, the idea of sustainable development has provided a philosophical base for delivering the future.

This became clear at the 1992 Earth Summit in Rio. The Earth Summit was in many ways the culmination of the work that began at the UN Conference on the Human Environment in Stockholm in 1972, from which came the UN Environmental Program and, later, the work of the Brundtland Commission. The nations that met in Rio committed themselves to sustainable development as a principle for future national and international actions, and so set the international environmental agenda for the next century.

Sustainable development will change the way we think about and interact with the environment. Therefore, it is important to understand where this idea came from and where it is leading us – and also to look at how existing green plans are adapting it to fill the gap between theory and practice.

A Brief History of Sustainable Development

Although Thoreau, Marsh, and Muir were concerned about ecological sustainability a century earlier, current thinking about the subject has its roots in the 1960s and 1970s, when the exponential growth of human population and the pressures it was putting on the natural environment were causing alarm. People began to understand that the world has finite resources and a finite ability to absorb the effects of our activities. Some scientists were concerned that we might fast be approaching the limits of what the earth could support.

Reports began to appear detailing the results of scientific studies on the ultimate effects of growth on the planet. Some proposed that humans, in addition to adopting population control measures, needed to change their economic activities if we were to avoid environmental and economic disaster. Some worried that our poor understanding of the effects of our activities, and the inevitable time lag in responding to environmental crises, would cause us to do permanent damage to the earth's carrying capacity.

All of these early studies urged some form of rapid self-imposed limits to growth, both in population and in economic activity. The 1972 MIT study commissioned by the Club of Rome, *Limits to Growth,* recommended a number of measures designed to achieve an "equilibrium state" between resources, population, production, and consumption.[2] Paul and Anne Ehrlich, in their book *Population, Resources, Environment,* recommended "de-development" for developed countries and a new type of semi-development for underdeveloped countries.[3] All assumed that a more equitable distribution of wealth and resources between developed and underdeveloped countries would have to occur.

When it was published in 1987, the Brundtland Report acknowledged that the world could not support the continued growth of *current* economic practices, but rejected the idea that growth itself is necessarily unsustainable. It argued that a certain amount of growth is essential, particularly in the developing nations, and that the integration of environmental and economic considerations in human decision-making processes could lead to greater efficiencies in resource use and to intelligent, equitable economic growth.

While few challenge the Brundtland Report's assessment that development must occur and standards of living must be raised in underdeveloped nations, there is still significant opposition to the idea that quantitative economic growth can be sustainable. Economist Herman Daly, one of the premier voices in the growth debate, argues that, because the natural world we live in has physical limits, so must the physical dimensions of our production and consumption of goods. Daly supports the concept of sustainable development, but makes a clear distinction between growth and development, arguing that "'growth' should refer to quantitative expansion in the scale of the physical dimensions of the economic system, while 'development' should refer to the qualitative change of a physically nongrowing economic system in dynamic equilibrium with the environment."[4]

Daly argues throughout his writings that continued economic growth does not necessarily contribute to people's well-being, economic or otherwise, while it inevitably diminishes the earth's "natural capital," or ability to provide humans with goods and services. He contends that current economic theory is based on a false picture of what humans are and how they act, an abstraction that relies too heavily on market transactions and income as measures of well-being and denies the primary importance of social relationships in human welfare.

21

Figure 1

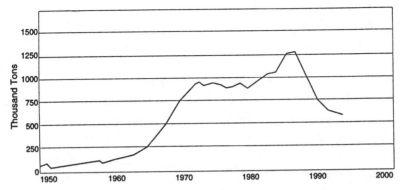

1. World Production of Chlorofluorocarbons, 1950–93. In addition to consuming more than their share of resources, the industrialized nations produce approximately 80 percent of all CFCs, which are responsible for damaging earth's ozone layer. Between 1950 and the mid-1980s, global production of CFCs went from fewer than 50,000 tons per year to more than a million. Source: Worldwatch Institute, *Vital Signs 1994*, based on data from the Chemical Manufacturers Association and E. I. Du Pont de Nemours.

Daly's theories have not been welcomed by traditional economists, but his work, and that of others such as Hazel Henderson, has had some effect on the international policy debate. The question of whether economic growth can be sustainable is likely to be argued far into the future, along with another, closely related, question raised within the context of sustainable development: how a more equitable distribution of resource use can be achieved.

The idea of moving toward equity in resource consumption is very important to Brundtland's and others' definitions of sustainable development, most of which attempt to resolve the historic problem of industrialized nations appropriating the resources of southern hemisphere nations. The current reality is that the industrialized nations consume far more than their share of resources. According to the Worldwatch Institute's 1994 *State of the World* report, in 1989 the richest 20 percent of the world's people absorbed almost 83 percent of global income, while the poorest 20 percent received only 1.4 percent.[5] The wealthiest 20 percent have consumed energy, raw materials, and manufactured goods at a correspondingly high rate.

The equity issue is further complicated by the fact that the earth's diminishing pool of resources simply cannot support such high resource consumption in the underdeveloped nations or by future generations. Unless the industrialized nations cut back their use of resources, others cannot hope for much improvement in their standards of living. It is clear that if we continue to follow current practices, we will create an environmental legacy that will reduce the standard of living for *everyone* in the future, and cause increasing human conflict.

The concept of sustainable development defines the problem, but what are the solutions? How does one nation adopt a policy or policies that adequately address issues of a global nature? A nation's answers will differ greatly depending on whether it is developed or developing.

Much of the discussion of sustainable development in the industrialized nations has focused outward, toward the unsustainable practices of developing countries, many of which are driven by poverty and the desperate need for economic growth. But many in the developed countries tend to forget that their own practices are an equal part of the equation. Developed countries not only consume the most resources, but are also the source of most environmental problems (see fig. 1).

Developed countries need to get their own houses in order, adopting rational, comprehensive policies that will make their resource use more efficient and clean up their messes, before they can expect to be very effective at telling others how they should manage their affairs. Developing countries are not going to respect demands that they take better care of their resources unless developed nations first show that they are serious about doing their part. Until this happens, the underdeveloped countries have every reason to be apprehensive about either the intent of the developed nations or their intelligence.

This does not mean that developed countries should ignore global questions. Pursuing courses toward the sustainable use of their own resources does not mean that they will stop pursuing international agreements or providing financial and technical aid to the developing countries. The developed nations must also be prepared to share information and technology relevant to sustainability.

Developing countries will benefit in any case from this approach. New efficiencies practiced by developed nations will free up resources for use in the developing world. In addition, the pioneering work of industrialized countries will develop a body of experience that should prove useful to the underdeveloped nations as they begin to create their own plans. It will allow them to study and evaluate what has already been done, and to think about which approaches might be best for their own planning.

Brundtland's concept of sustainable development is a philosophical statement, not an action plan – though it has been misinterpreted as such. The governments that have developed green plans understand the distinction well; while they have adopted the principle of sustainable development as their ultimate goal, they have also forged ahead with more refined definitions and action plans regarding the sustainability of their own resource base. By applying the concept of sustainable development in the real world, they are defining it as they go.

The rest of this chapter will look at how each of the three main green plan countries are dealing with the sustainability question on a practical level. Each has taken a somewhat different approach, tailored to fit differing national environmental problems and social characteristics. New Zealand's approach is primarily structural, rethinking its environmental laws and governmental structure so that both are now directed toward the principle of sustainability. The Netherlands, with its higher level of environmental problems, has chosen a technical interpretation of sustainability, aiming to recover a high level of environmental quality within twenty-five years. Finally, the Canadians use the concept of sustainable development as a theoretical framework that helps order their thinking about environmental issues and decision-making processes. They have also found it a useful way to keep people thinking about the big picture, instead of getting bogged down in the parts.

New Zealand's "Sustainable Management"

When New Zealand was going through the process of creating a new resource management policy, there was a thoughtful and lengthy debate in Parliament over the real meaning of sustainable development. Unable to reach any agreement, the legislators eventually realized it was more important to define what they wanted to achieve in their country, which was resource management that would be sustainable over time.

They adopted the term "sustainable management" to describe what they wanted their new legislation to accomplish. Their definition of the term can be summarized as "managing the use, development and protection of natural and physical resources in a way, or at a rate, which enables people to meet their needs now without compromising the ability of future generations to meet their own needs."[6]

Sustainable management, as the New Zealanders see it, is their route to achieving sustainable development as defined by Brundtland, but is a more

24

accurate description of how they will actually apply it to their own environment and economy. They are well aware of the international responsibilities of sustainable development, but know that international goals cannot be achieved without first meeting national ones. They are cleaning their own house first.

Decision making within New Zealand's concept of sustainable management involves a scale of environmental protection. At one end are environmental "bottom lines," which are points beyond which there is an unacceptable risk that a resource or ecosystem will suffer irreversible effects.[7] These bottom lines represent the minimum of environmental protection in New Zealand; ideally, no decisions will be made that would allow a system to go below this line.

On the other end of the scale are points beyond which no modification occurs whatsoever, cases in which ecosystems or areas will effectively be set aside and no direct human influence allowed. In between these two are areas in which decision makers will have to weigh environmental, economic, and social concerns. Within these parameters they hope to establish standards that will define sustainability across the environmental spectrum.

Sustainable management is neither anti- nor pro-development; it is only concerned with the effects and potential effects of development on the environment's sustainability. But the New Zealanders believe that a distinction must not be made between development and the environment; in other words, development must fit into the framework of sustainable management of resources. Although they think it is possible to have development and growth, they realize that there must be conditions placed upon it to ensure sustainability.

The Netherlands' Technical Approach

The Dutch spend very little time talking about sustainable development, but it is nonetheless the guiding principle of their environmental policies. Although they had been moving in the direction of a more comprehensive environmental policy before the Brundtland Report was released, their green plan, the National Environmental Policy Plan (NEPP), was to a large extent a response to Brundtland.

The Dutch have chosen to focus on Brundtland's appeal that nations stop shifting responsibility for environmental problems onto future generations and other countries. That is why they have set themselves the goal of recovering the Netherlands' environmental quality in one generation, and are not allowing themselves to reach that goal by, for instance, shipping their wastes to other nations. They also use the idea of equity for future generations to inspire their people to make environmentally sound decisions.

Overall, the Dutch define sustainability for their own country in terms of reducing emissions and reversing unacceptable practices in each of their environmental problem areas (what they call "themes") to such an extent that the environment can continue to function at an acceptable level in the future. With the help of independent scientists, they have already determined the reductions required in each problem area and for each "target group" that contributes to those problems. They are in the process of developing sustainability indicators – instruments to help them track their progress – for all the themes.[8]

Sustainability, the Dutch believe, must be pursued by "feedback mechanisms" aimed at the sources of environmental deterioration. These feedback mechanisms focus on the proper management of material and energy flows. They are: integrated cycle management (closing material cycles in the chain from raw material to production process to product to waste, in order to eliminate leaks); quality promotion (improving the quality, rather than increasing the quantity, of raw materials, production processes, and products); and reduction of energy use.[9] Attention must be focused on all three of these elements simultaneously, the Dutch believe, if sustainability is to be achieved.

It is interesting to note that, in the four-year report on their plan, they anticipate that even more fundamental changes may be necessary if they are actually going to achieve sustainability:

> We are on the right road, but nonetheless the figures presented in this Environmental Outlook show that our efforts to date have been insufficient.... As long as the material growth of production and consumption continues, extra efforts will be continually needed in order to stay within the environmental constraints.... As we run out of technological options for change within the limited time available for such measures, this will force a fundamental revision of our expectations about the nature and the extent of "economic" growth.[10]

A Useful Tool for Canada

Like the New Zealanders, the Canadians struggled with the concept of sustainable development early in their green plan process. They also discovered how difficult it is to come up with a definition that is precise and widely accepted, but learned that they could narrow it in ways that could be useful to them.

They now use the term as a general statement of goals, a theoretical framework within which they can develop an actual program for action. In contrast to New Zealand and the Netherlands, Canada has not tried to apply sustainability to its own resource base by developing environmental bottom lines or precise standards. Rather, it sees sustainable development as a tool for organizing its thinking about environmental issues.[11]

Like the Dutch and New Zealanders, Canadians believe that sustainable development requires an understanding of people as well as ecosystems, and the interactions between them. In particular, they believe that it is necessary to study how and why humans make decisions, so that environmental values can be integrated into all their actions, as employees and employers, as consumers, as family members, and as members of their community. In their vision, this is the true promise of sustainable development: that if they are able to change behavior, they can grow economically *and* maintain and improve the environment. However, their concept of development is not restricted to economic growth.

The Canadians also see the principle of sustainable development as a good way to get people working together to solve environmental problems. They emphasize the idea's inherent optimism – that it is possible to make good decisions, take positive action, and solve these problems, if we just put our minds to it. It is a fundamentally positive way of approaching the challenge of environmental recovery, and this optimism runs throughout the Canadian green plan.

They have also pointed out that it is useful to be purposefully ambiguous in the definition of sustainable development: if too much attention is paid to specifics when developing an overall strategy, there is a risk that people will immediately fasten on just one issue or part of it. There is a tremendous advantage in using a big-picture approach instead of loading the definition with detailed statistics, because the idea of thinking comprehensively is simply too new; people have to learn how to do it.

The big picture of sustainable development can help people understand that it is a long-term process. If they do not, they will see only a snapshot of the program as it exists at the moment, and will peck away endlessly at one point or issue. All these countries' plans will inevitably change; they will discover new directions through the experience and improve as they go along. The Canadians realize that it is important to keep their plan framed within the large-scale, long-term context of sustainable development, so that it can be properly evaluated.

Rio and Beyond

The experiences of the New Zealanders, Canadians, and Dutch show that the idea of sustainable development can be adapted and refined in various ways to make it a useful theoretical tool, from the national level on down to the local. The theme of sustainability will be repeated in one form or another as both developed and developing nations create green plans.

And in the international community, where sustainable development has been discussed and debated ever since the Brundtland Report, the idea is gaining ground. At the 1992 Earth Summit in Rio, more than 170 nations agreed to Agenda 21, which promotes cooperation on environmental recovery through the principles of sustainable development. Like the green plan idea, Agenda 21 grew out of the Brundtland Report and its concept of sustainable development. Thus it is no surprise that the two share some of the same principles and management strategies, the most important of which are the integration of economics and environment into decision making on all levels, and a comprehensive approach to resource management.

Agenda 21 is not itself a green plan, but rather an agreement on sustainability as a common goal for the nations of the world, and a framework for international cooperation toward it. In that context, it urges all countries to adopt national strategies for sustainable development – in essence, green plans. In addition, the UN Development Program has announced a new department within the UN that will aid developing countries in implementing green plans of their own.

Agenda 21 is already having major impact in the world. For example, the annual report of New Zealand's Ministry for Environment to its house of representatives declares that one of the Ministry's objectives is to work with other departments, local governments, industry, and community groups to ensure that the programs of Agenda 21 are incorporated into making policy and decisions. This is one of the first nations to move ahead with fulfilling its obligations under the agreement.

Increasingly one is seeing Agenda 21s on the local level. The Rio agreement particularly emphasized the desirability of having communities develop their own comprehensive strategies for managing their affairs sustainably.

The excitement of Rio is still reverberating around the world. Implementing Agenda 21 in all the signatory countries will take time, but it will happen. As the document's preamble states, "This process marks the beginning of a new global partnership for sustainable development."

There are many wonderful possibilities – and dangers – inherent in a concept as ambiguous as sustainable development. In his book *The Wealth of Nature,* Donald Worster argues that the idea has deep flaws: first, that it is based on the idea that the natural world exists only to serve humans; second, that it assumes that we will be able to determine the carrying capacity of ecosystems, when if we have learned anything about nature, it is how unpredictable it is; and third, that it assumes sustainability, whatever that might be, can be achieved without significantly altering our current values and institutions.[12] Worster is also concerned that economics will drive sustainable development, while environmental issues trail behind. "Sustainability is, by and large," he argues, "an economic concept on which economists are clear and ecologists are muddled."[13]

Others have noted the same weakness in the concept. The Dutch Committee for Long Term Environmental Policy has pointed out that, as it is currently understood, sustainable development leaves "too much room for other interests besides the environment," and adds that it is crucial that more emphasis be put on "sustainable environmental quality" in the future.[14]

These are important criticisms and caveats, and they need to be thoroughly considered as we refine what we mean by sustainable development and begin to apply it. However, as noted earlier, sustainable development is not an action plan but a philosophical statement, a way of thinking about how we relate to the natural world. If it is vague and overly broad, it has also helped us approach consensus on some very important issues. I cannot agree with Worster that sustainable development lacks "any new core idea"; in fact, I believe that it gives us a vision of new ways to interact with the natural environment. How we apply that vision is up to us.

• • •

What does sustainable development really mean for the future of the world? It all comes down to this: The earth has a certain capacity, and to be able to manage it in an environmentally as well as economically friendly way, we will in the end have to make some choices about how much we are going to produce and how much we are going to consume, and about the ways in which we produce and consume. That also means dealing with a related issue – one that is hard to discuss even in the United States – population growth.

When exponential population growth and the poverty it is linked to were confined to the underdeveloped world, the developed nations believed they could ignore the issue. But another outgrowth of both overpopulation and poverty, mass migration, has made this increasingly difficult. Mass migrations of people from the underdeveloped world are becoming more and more common, and as the world grows smaller, travel will become easier even for those who have no money. There will be growing pressure on the western world from the south and from eastern countries in Europe.

Mass migration not only overburdens the areas people are moving to, but often further impoverishes the areas they came from as well. The solution most often overlooked is to make social and economic life better for the citizens of countries at the source of the migration. In order to do this, the developed nations will have to keep a tight rein on the size of their own populations and become much more efficient in their use of resources, because each citizen of a developed nation consumes a much higher proportion of resources. Developed countries will also have to help the developing ones improve their use of resources while at the same time lessening the pressure on the environment. All this will require greater resource efficiency and different ways of producing and consuming.

In the long term, it is in the developed nations' best interest to help developing countries adopt their own plans for sustainability, because only through this process can a solution be found to the problems of poverty, overpopulation, migration, and environmental decline. In this era of interdependency and shrinking distances, the developed and developing worlds are inextricably linked.

The process that began in Rio need not be a frightening or antagonistic one. It is an opportunity for people around the world to come together and work toward a livable future, within their own nations and on the international level. Green plans, with their promise of ideas, experiences, and knowledge to be shared among nations, provide just such an opportunity.

3

A Green Plan Predecessor:
California's IFP Program

My own experiences in government laid the foundation for my firm belief in the need for a comprehensive, integrated form of resource management. There I saw firsthand the sorts of problems that can arise when we attempt to manage the environment in the traditional way, dealing with single-purpose issues one at a time, especially when confronted by a sudden, massive, complex environmental challenge.

From 1977 to 1982, I was head of California's Resources Agency. This is the state-level version of a minister of environment. The department's budget at that time was nearly $1 billion, and it had fourteen thousand employees. As a member of Governor Edmund G. (Jerry) Brown Jr.'s cabinet, I was responsible for the administration's resource policies.

This was no small responsibility, because California is heavily resource-dependent. Tourism, agriculture, and timber production are three of its largest industries, and its fisheries have traditionally also been important, although they have suffered badly in recent years as wild fish populations have declined. In addition, California is a heavy consumer of energy and water for agricultural, industrial, and residential uses.

When I took the job, my goal was to promote the idea of stewardship of resources as a public trust. I felt that, in California and the nation as a whole, we had thought only of harvesting and consuming the cornucopia of our natural resources. We had failed to understand, or chosen to ignore, the fact that these natural systems require sensitive management to keep them healthy. I did not set out to develop a comprehensive, integrated plan for dealing with California's resources; I realized the importance of such a plan only when forced to confront the pressures that the energy crisis of the 1970s put on the state and its resources.

Energy was a relatively new issue in those days, although it had been brought home with great force by the Arab oil embargo of the early 1970s, when the price of oil went up more than 150 percent[1] and sent energy prices through the roof. That increase was a threat to the state's entire economy, and the resources agency naturally spent a great deal of time reacting to the fallout from the crisis.

A great many people believed they had the answer to the problem, and the advocates of nuclear power were some of the loudest among them. I was against the development of nuclear power, primarily because there are no reasonable solutions for dealing with the waste. One of the things I did when I first took office, then, was to put the agency in opposition to the plans for more nuclear plants. With the governor's support, the resources agency announced it was going to shut down nuclear power.

California was already facing a serious threat to the economy from the energy crisis, so that decision caused a tremendous public relations furor. The utilities fired back that they had to have an energy growth rate of 7 percent per year forever, and that California would shiver in the dark if its nuclear power generation was not increased. Some years later, the then-president of the largest utility, not just in California but in the world, publicly thanked us for taking a stand against the industry's plans to build forty nuclear plants up and down the state. If his utility had carried out its plans, he said, it would have gone bankrupt.

In Search of a Comprehensive Solution

But stopping nuclear power development still left us with the problems caused by the energy crisis. As the angry calls poured into the agency asking what we were going to do about it, we struggled to come up with an answer, to devise an energy policy that would make sense for the future. One night I literally awoke to the realization that this issue affected every other issue my department dealt with.

For instance, one of the responsibilities of the resources agency is water delivery throughout the state. The largest users of energy in California are the water pumps, which run constantly. The pumps perform a range of functions, from keeping up pressure in the mains so that water comes out when someone turns on a faucet, to pumping water more than 600 miles (including up and over a mountain range) from northern California to Los Angeles and the agricultural and residential users of the arid south.

This realization led to understanding that water delivery, energy consumption, and air pollution were connected. A portion of the state's air-quality problems come from burning oil in order to create the energy to pump the water. And then we began to realize that the poor air quality was somehow affecting trees and crops; we did not understand how or why, but our suspicions were later confirmed when the problems of acid rain were better understood.

All that is to say that the interactions of humans and their environment need to be dealt with in a comprehensive way, because they reach further than one initially realizes, and are all interconnected. Faced with the far-reaching crisis in energy, that understanding struck me with great force. At that point I began to develop plans for a more comprehensive resource strategy, which came to be called Investing For Prosperity (IFP).

The philosophy behind IFP was this: nature has its own time scale. It takes one hundred years to make an inch of productive topsoil, a half-century to produce commercial timber, and decades to restore degraded range lands. If we tie our response to the erosion of our resources to annual government budgets, fought over from year to year and from crisis to crisis, we will never succeed. Until we have a comprehensive and integrated plan, we cannot move forward.

And California had many resource problems, even beyond the immediate ones caused by the energy crisis, that called out for such a comprehensive plan. The state's timber production had been declining for two decades, to the point where commercial timberlands were producing less than 50 percent of their capacity. Salmon runs were dropping precipitously, and the productivity of thousands of acres of farmland was threatened by erosion and increasing soil salinity.[2]

The idea behind IFP was to establish funding for and set up programs to restore and improve many aspects of the state's natural resources. The resources agency was able to show that the returns to the state would be well worth the investment. Eventually we got a series of laws passed that provided more than $120 million a year for investment in long-term quality and productivity programs for California's natural resources.

The relevant fact for this book and for the green plan idea is that these programs have been in existence for fifteen years; where other comprehensive strategies are too new to assess very accurately, this one has measurable results. More than a billion dollars has been invested in the IFP programs, and they have returned millions on the investment. Taking a close look at them is one way to evaluate the potential success of the green plan idea.

For instance, in the energy area alone we achieved some remarkable results. We invested $48 million in energy conservation and alternative sources. By 1988, energy use at state facilities had been cut by 20 percent; our goal was a 40 percent cut by the year 2000. Our goal for energy development of alternative sources such as cogeneration, geothermal, solar, and biomass was four hundred megawatts; by 1990 we had achieved 191. By 1990, the combined savings to the state from these energy programs was about $333 million.[3] Despite the fact that we did not reach our initial goals, the savings we achieved were quite substantial. In addition, our efforts inspired industry to make a massive turn toward conservation, and businesses soon took the lead on energy efficiency.

There were individuals and departments other than the resources agency that helped make this investment in energy alternatives a reality, of course: the State Energy Commission, for instance, and the Office for Appropriate Technology, as well as certain committed individuals. The California legislature, which responded to the challenge of the energy issue early on, was also a major player.

The success of the energy component of IFP was also due to the support of individuals and organizations outside of government. Examples include the corporate leaders who adopted the idea and ran with it, the managers of institutions such as hospitals and schools who did the same, and everybody right on down to the people who turned off the lights in their closets. These were the people who started an energy revolution in California.

Unfortunately, we have not followed through with the same enthusiasm on other issues, be they water, soil, wildlife, or air. But there have been changes in these areas, too – and there will be more over time – that may one day look just as good as our successes in energy conservation. At least better policies are in place for air quality, and they are starting to fall in place for water.

The Fight for Passage

Because time was of the essence in getting the IFP legislation passed, the resources agency did not conduct public hearings or field hearings. That would have been ideal in terms of building a strong constituency for the programs, and is what New Zealand did for its Resource Management Act. But the conditions we faced were difficult. Just the decision to stop nuclear development was creating an uproar of a dimension I could not have imagined, with attacks on all sides by the public relations firms the utilities had hired.

Worse, the state was in the throes of its taxpayer rebellion, which would eventually lead to Proposition 13 and its severe curtailment of government revenues from property taxes. The order went out that there would be no new programs; government spending had to be cut. I believed that, instead of being a negative, the Proposition 13 crisis had to be the catalyst for a new attitude toward resource management. The environment had to be seen as an investment, not just another expense, because ultimately it is these resources upon which the economy and the quality of life are based. But I knew that it would take quick and aggressive action to win my point.

Fortunately, I was able to move aggressively because I had trusted allies, the people I had brought in to head the various departments under me: forestry, fish and game, water, and so on. I asked them to find the dreamers in their departments – those who were the brightest and most able, but also the most frustrated – and pull together all their ideas for improving the departments' effectiveness.

Once they had done this we had several meetings to discuss the results, and within a few weeks I was able to lay out what would become the IFP fund. I then had to go and sell it statewide, and proceeded to do so with the help of those departments, which now had a stake in the issue.

We signed on constituency groups like the League of Women Voters, which adopted the resource fund idea as its statewide project for the year. The Audubon Society was the first environmental group to support IFP, followed by all the others. Eventually, even the labor unions and corporations, including heavyweights like IBM, Southern Pacific Railroad, and the Bank of America, came to understand what we were doing and openly supported us.

One example of how we were able to forge such a strong coalition is the approach we took to the forestry industry. I went directly to the main lobbying organization for the entire forestry industry, which would typically have opposed the agency on this issue. I met with the director and laid out the plan, and he said he would check with his constituents and get back to me. When he did, he said, "I'll surprise you: they're willing to cooperate, but

35

they have one requirement."

That requirement turned out to be an interesting challenge. The forestry industries' concern was that about one-third of California's timberland was not producing. One reason was that the small, private owners and investors could not afford to replant their land and then wait fifty years for a new crop of trees. If I could figure out something that would help the small owners to get trees in the ground and make their lands productive, then the forestry industry, would help me on the big picture. That included lobbying in the areas of energy efficiency, water, and whatever else it would take to pass this bill for $125 million a year in resource investments.

So the agency created a grant program for small timber owners. In order to get a grant, though, the owners had to put together a productivity plan, prepared by a professional forester, that would lay out the long-term future for the lands involved. They could then apply for a grant to do whatever needed to be done, whether it was building bridges, improving their roads, planting trees, or preparing forests.

Jeff Romm of the University of California's Department of Forestry has since said that the most important part of this program was its requirement of stewardship, that the owners had to put together a management plan for the future.[4] It changed their attitude and outlook, he said, so that henceforth in their meetings, individually or in the professional associations, they would talk about the long term; they would no longer just talk about getting board feet off the land. Although the program started out primarily as a push to get trees planted, the philosophical point has proved more powerful.

The lesson to learn from the timber growers is that everyone has something to gain from improving the quality and productivity of the environment. This appeal to a wide variety of interests is one of the strengths of a comprehensive program. Some interests will be opposed to parts of it, but because many more will benefit from it as a whole, the opposition will find it harder to block. Any state that goes through a green plan process will have to reach out to all the individual interest groups and build a powerful constituency for change.

Investing for Prosperity: The Program

Between 1978 and 1980, with coalition-building and other hard work, the resources agency was able to achieve passage of all five pieces of legislation that created the legal and financial basis for IFP. Taken together, they constituted a program of investment in California's natural resources that aimed to put the state back on healthy environmental and economic ground within twenty years, while looking ahead one hundred years to the legacy we would leave our grandchildren. Because of their success, many of IFP's programs continue to be funded, despite several changes in state government since then.

The five laws that established IFP were:

- the Forest Improvement Act, which provided for an urban forestry program and cost-sharing of reforestation on small private forest lands
- the Forest Resources Development Fund, which established the principle that income from the sale of state-owned timber would be reinvested to improve forest productivity and implement urban tree-planting programs
- the Renewable Resources Investment Fund, which provided $10 million to develop wood energy, help restore salmon stocks, and implement water conservation and reclamation projects
- the Geothermal Resources Fund, which provided that 30 percent of the income from federal geothermal leases in California be deposited in the Renewable Resources Investment Fund
- the Energy and Resources Fund, which allocated a portion of the state's tidelands oil revenues for the restoration and protection of the state's renewable resources[5]

Under the program, the funds made available by the legislation were to be invested in a number of resource areas. Specific goals were established for each, and projects implemented to achieve them. The main areas were forestry and wildlands, fish and wildlife, water, soils, coastal resources, parks and recreation, and energy.

In response to the energy crisis, we had the following goals for the year 2000: to reduce by 40 percent the amount of energy used by state government operations; to continue to expand and encourage efficiency; to develop efficient and renewable energy production technologies; and to reduce gasoline consumption by forty percent of the 1980 levels.[6]

We decided early on that there was no one black box, like nuclear power, that was going to be the answer to our energy problems. We learned that any small amount of alternative energy you could produce, even as little as a half a percent, would in time accumulate, together with the savings from conservation and the energy created by other alternative sources, into a big block of energy that could fulfill the state's needs. There was considerable pioneering in the development of alternatives to fossil fuels in those years, such as cogeneration, biomass, solar, and geothermal.

Some of the programs that the energy fund invested in were government focused, since we believed we had to show the way before we could expect the rest of society to follow. These projects included streetlight retrofits, replacing old light bulbs with more energy-efficient ones; conservation in schools and hospitals; and new appliance and building standards. The retrofit program for streetlights alone is saving cities some $2 million per year, while the $45.3 million that IFP loaned to schools and hospitals to develop cogeneration projects has resulted in a net savings of over $700 million.[7]

Other programs have also done very well. From 1979 to 1988, $30 million was invested in salmon and steelhead restoration programs, 90 percent on the North Coast and 10 percent in the Central Valley. Five hundred and twenty-three restoration contracts were funded, about half of what we had originally hoped to fund.[8] However, we surpassed our original goal of clearing the blockages from five hundred miles of streams so that salmon could swim up into them and spawn freely; by 1988, fourteen hundred miles had been cleared.[9]

It is difficult to assess the impact of a program like the salmon and steelhead restoration in the short term, because other factors cause fluctuations in the results. In this instance there was one huge blip upwards, after which the fisheries again began to decline. Much of this has been blamed on high-seas netting, in which Korean, Taiwanese, and Japanese fishermen were creating walls of drift nets out in the ocean that trapped anything that swam, including salmon and steelhead. (This has now been changed through international treaties.)

The low stream flows caused by the drought conditions in California during the years 1987 to 1992 have also caused many problems for the fisheries. With the sudden return of heavy rains in 1993, and with several more years of adequate flows, we have hopes of bringing them back. In 1994, Oregon and Washington were required to halt all salmon fishing due to the decline in wild populations. California's salmon fishing season remained open, however; in fact, it was the best in many years. Fishing industry representatives believe that the improvement in salmon populations off the California coast is largely due to the work done under IFP. It is clear that the stream-clearing program was well worth the investment: without it, conditions now would certainly be worse.

In the forestry sector, part of the plan was to help reforest a million acres with 360 million trees to reverse the decline in timber productivity. A study funded by a subsequent governor's administration projected that, over the next fifty to seventy-five years, the $5 million spent on this project will return $448 million in timber sales, $104 million in tax revenues, and $9 million in consumer savings, while creating eighteen thousand new jobs.[10]

IFP also had a goal of increasing the annual timber supply over 1980 levels by three billion board feet through improved wood products utilization, integrated pest control, and forest and tree improvement efforts. That demonstrates the program's comprehensive approach: instead of investment in just one thing, growing trees, it promoted such innovations as using thinner saw blades in the mills because they caused less waste than the old ones. It is the same principle of conservation that worked so well with energy.

IFP proposed some changes in agriculture that were designed to conserve soil, particularly in terms of research into reducing the consumption of fuel and other oil by-products. Other ideas included minimum tillage, windbreaks, and integrated pest management – whatever one could do to cut down on the use of chemicals and tractors.

IFP did not ignore the state's urban resources, either. Substantial funding went to park development, beach erosion control, and reestablishing beaches in cities where they had been lost. It was this that convinced the labor movement to come along with IFP. Labor unions understood the energy issue, but were not terribly excited about it. However, the idea of enhancing the quality of life for workers by investing in outdoor recreational opportunities did appeal to them. They liked the fact that we emphasized trail development, campground development, and other themes in the parks and recreation area.

Because water is such a scarce and precious resource in California, we also devoted a great deal of attention to water conservation projects for both agricultural and urban users. These included investing in a computerized system for agricultural irrigation that was based on actual crop needs rather than farmers' traditional methods. Urban water conservation practices were encouraged by the distribution of more than four million kits to households throughout the state. This program saved 38,700 acre-feet of water and 940 million kilowatts of energy per year.[11] A subsequent conservation program cut urban water use by one-sixth in a short period of time.

Changing Government

In the process of getting IFP passed and implemented, I learned many valuable lessons about what it takes to make a major change in government. One such lesson was how to get a program funded when it seemed impossible.

In spite of California's taxpayer revolt, we were successful in obtaining funding for this idea because we wove all the elements together into one big picture, in a way that reached industries and labor unions and environmental groups and educators and scientists. They supported it because they could see the vision behind it, and because all felt they had a part in it. The green plan creators in New Zealand, the Netherlands, and Canada have all had the same experience.

On a pragmatic level, the experience also taught me how important it was to get all the relevant government agencies working together. For instance, on my first day on the job I found that the energy commission, whose budget I was responsible for, was suing the forestry department, another of the agencies under my authority, because of a conflict over water and energy issues. Neither of the two agencies had bothered to contact my predecessor.

When government agencies and policies are not integrated in a comprehensive way, it leads to this kind of wasteful squabbling and creates tremendous inefficiencies. When the public sees this going on it loses faith in government, as was happening in this case. We could never have accomplished what we did if we had allowed that situation to continue.

The most important factor in IFP's success is that we made it our top priority, and the governor was also committed to the idea. Leadership from the executive and managerial branches can be crucial through the inevitable political ups and downs. A program like this can only be maintained if it has been firmly planted.

The next administration tried its best to kill what we had started. The new governor did not think much of our ideas; I believe he asserted that there is "no such thing as limits" in regard to resources. The energy crisis dissipated, and the new governor saw his role solely as one of creating jobs through growth.

But those people who were committed to the program and stayed on in government were able to sustain it at about 80 percent of its original levels. The sound energy policy laid out by IFP is fairly standard practice now, regardless of who is in office. Other individuals who believed strongly enough in the idea have kept it alive elsewhere.

●　●　●

The program and its goals caused a tremendous uproar, played out in the California legislature between the environmental movement and the utilities and other industries. But once the smoke had cleared, we had achieved something that benefited the whole state. It took a large and diverse coalition to accomplish it, but that coalition became one of its greatest strengths.

PART TWO

Assessing Green Plans in Action

4

The Netherlands: "Each Generation Cleans Up"

The Netherlands' green plan, or National Environmental Policy Plan (NEPP), is the best job of technical environmental planning done by any nation to date. Its strength is due to the careful thought put into it by hundreds of the country's brightest people, who have managed for the first time ever to put complex, interrelated natural systems into a manageable context. Not only does the NEPP successfully weave together complex systems such as water, air, soil, and energy; it also meshes them with human factors of economics, health, and carrying capacity. It is a comprehensive, long-term, well-thought-out plan, and the Dutch are putting billions of guilders into implementing it.

Equally important is the NEPP's emphasis on a strong management framework, which has helped to make it practical and functional. Traditional environmental programs have tended not to work well in the real world because they focused on one or two systems and were not involved with management overall. The Dutch were fortunate that Pieter Winsemius, who is a management specialist in his work outside of government, was the minister of the environment during the planning phase.

This emphasis on good management has led the Dutch to some interesting innovations, particularly in the way they deal with industry. One example is the way in which each industrial sector has been encouraged to organize into associations and to work out for itself, in cooperation with the government, its plans for achieving the environmental quality goals established in the NEPP.

Another is the twenty-five-year time frame they have set for achieving environmental recovery. The Dutch chose twenty-five years, or one generation, as a framework. They use the phrase "each generation cleans up" to express this principle as a clear, easily understandable goal that every citizen can support.

The Netherlands has a long tradition as one of the world's most progressive nations. It has maintained literacy at some of the highest levels in the world, and has always been in the forefront with its policies on quality-of-life issues such as health care. In creating and implementing the NEPP, the Dutch have once again taken the lead with a program that can actually solve the environmental problems that are of such concern to every nation.

The Netherlands' Environment

There are many good reasons why the Netherlands became the first nation to adopt a comprehensive national environmental strategy. It has always been a nation of limited natural resources, and has had to trade with other nations to survive. In addition, the Netherlands today is under a great deal of environmental stress. While every developed nation has suffered environmental decline because of industrialization, the Netherlands has many complex problems packed into a very small, heavily populated area.

The Netherlands is situated at the mouth of the Rhine, one of Europe's most polluted rivers. Every year the river deposits enormous quantities of contaminated silt at its mouth near Rotterdam. Air pollution also travels across Europe, so that the Netherlands suffers from the emissions of other nations' industries (and vice versa). Almost half of the acid deposition in the Netherlands originates in other countries, particularly the surrounding nations of Germany, France, and Belgium.

The Netherlands' economy is also heavily dependent on polluting industries such as commodity chemicals, metals, and agriculture. Agriculture contributes heavily to just about every environmental problem. Their successful agricultural export program has been based on a policy of intensive management and growth, and one of the results has been that the number of cattle in the Netherlands has doubled since 1970, putting even more pressure on their agricultural lands and water resources (see fig. 2).

The Dutch have about 1,145 people per square mile, compared to about seventy in the United States.[1] If you include about 40 million pigs, 100 million chickens, 7 million cows and 6 million cars,[2] you can imagine the enormous amount of pressure being put on the environment. The impact of the growing number of people and animals on this small European landscape was one factor pushing them toward comprehensive environmental planning.

Figure 2

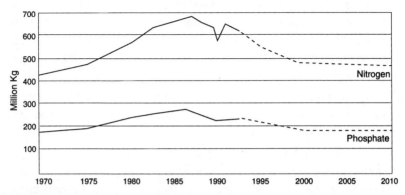

2. Phosphate and Nitrogen Content of Manure Production in the Netherlands, 1970–2010. The Dutch practice of intensive livestock farming produces more manure than farmers can dispose of properly; phosphate and nitrogen in the manure contribute to the problem of eutrophication. The reduction in phosphates and nitrogen indicated by this chart follows a general reduction in manure production in the Netherlands since 1984. Source: The Netherlands' National Institute of Public Health and Environmental Protection.

But concern over the degradation of their environment was not the only reason the Dutch became leaders in the field of environmental planning. Planning is not only familiar to them, but something they welcome. Their huge system of dikes and windmills – of water *quantity* management – has always been close to the country's heart. They have reclaimed enormous amounts of land from the sea in a period of about three centuries, and the process still continues. (This fact also makes them very aware of the dangers of global warming, because rising seas would put them in a desperate position.) The Netherlands as we know it simply would not exist without a great deal of intricate planning.

Finally, the Dutch are survivors. They have managed to be economically successful for centuries even though they have very few natural resources. While nations like the United States and Canada had huge forests and other resources they could exploit for an easy economic base, the Dutch had to learn to be careful, skilled traders and planners in order to survive. As a nation, they value cooperation and consensus in the realm of public affairs. The fact that they have been so successful at it gives hope for the future.

The Political Basis of the NEPP

Given this background, it is no great surprise that the Dutch government was studying more comprehensive approaches to environmental planning as early as the early 1980s. However, progress on new policies was rather slow, and there was little momentum for radical change until the watershed year of 1988. In September of that year, Queen Beatrix gave her annual state address, written, as usual, by the staff of the prime minister, Christian Democrat Ruud Lubbers. This speech had the Queen declare that the Netherlands' environment was actually improving – a claim that was clearly not consistent with reality, and which raised many eyebrows. One of the first things that happened in the next parliament was that Lubbers was asked for data to back up that claim, and those on his staff who had written the speech were unable to provide it.

Then came *Concern for Tomorrow*, a study outlining the seriousness of environmental problems in the Netherlands. Prepared by the National Institute for Health and the Environment (RIVM), a highly respected independent research foundation, the report quickly gained the public's attention, becoming a major media event. It predicted irreversible ecological damage to the country if the current environmental policies were pursued.

Concern for Tomorrow detailed what was wrong with the environment and what needed to happen to solve the problems. It looked at three scenarios for the future: one that would simply be a continuation of current policy, one that would involve only moderately stronger measures, and one that would solve the problems as we know them today. It concluded that, from the scientific point of view, the country should attempt the third scenario.

The final dramatic occurrence of 1988 came with Queen Beatrix' annual Christmas message to the people, which she herself writes. In this one speech each year, the Queen tries to touch upon issues close to her own heart, and this year she seized on the conclusions of *Concern for Tomorrow*. "The earth is slowly dying, and the inconceivable – the end of life itself – is actually becoming conceivable," she said, appealing for immediate action from the nation.[3] The Dutch, already shocked by *Concern for Tomorrow*, were galvanized by the Queen's speech. She became the spokesperson for the public's feelings about the environment, and has been dubbed the "Green Queen."

Because of the enormous environmental pressure on the Netherlands, every citizen understands the magnitude of the problem. They see that the few fish left in the Rhine are inedible, that the fees they pay for drinking water and waste collection are constantly going up, and that traffic jams have become a way of life as the number of cars continues to increase. Awareness is so high in the Netherlands that the citizens are willing to pay higher taxes in order to raise the money for environmental programs.

In addition to this broad popular support for environmental reform, the Netherlandic political parties, from right to left, were and are in agreement that the country needs a comprehensive plan. Because of this, the government had the political power it needed in order to make the plan work. These three factors – a public that was very responsive to the issue, a queen who was an inspiring leader and spokesperson, and support from parliament across all party boundaries – are the reasons the Dutch were able to pass the NEPP in 1989.

Public support for the plan itself appears to be very strong. In fact, it is now so deeply rooted in the Dutch political system that it is no longer controversial. That support has been evident ever since the events surrounding the first parliamentary vote on the plan, which caused the government to collapse – the first government in recorded history to fall over an environmental issue.

The collapse came about because parliament had selected, as a starting place for the plan, a proposal to end the subsidies that car commuters then enjoyed. The green plan advocates argued that if the goal was to get people out of their cars, it was essential to stop subsidizing driving to work. The proposal was voted down, the prime minister resigned, and a new government was formed. But when the question was put to the people, they voted overwhelmingly to pursue the NEPP on the scale it was intended.

Since that time, the country has reached such a high level of consensus about what should happen – how much money should be spent, how goals are to be achieved – that it is not likely to matter if a different leader or party takes over the government. Some individuals or parties might want to tinker with parts of the plan, but the essentials are agreed upon. People share the feeling that it is a good thing to do – good for the economy and good for the quality of life.

The NEPP legislation is titled "To Choose or to Lose," a title that was carefully selected in order to make its purpose clear to the Netherlands' citizens and to the world. It emphasizes the fact that humans have reached a point at which we must make some very tough choices if we are to survive. Our current policies will not suffice in the long term, although they may postpone disaster for a few more years. "To Choose or to Lose" means, then, that the Netherlands must choose a good long-term policy today if it does not want to lose out in the end.

In Amsterdam recently, a colleague of mine asked a cab driver what he thought of his country's national environmental plan. "Just another way for the government to increase the taxes on the poor working man," came the grumbling reply. Asked if it had no benefit for him, he answered, "Well, not so much, but it will for my children, and that is what is important."

The Philosophical Basis for the NEPP

The Netherlands' comprehensive, integrated approach to environmental planning did not happen overnight. Like most western countries, the Dutch first began passing environmental legislation in the early 1970s. This first wave of environmental law was devoted to single issues – regulating toxics, cleaning up wastes, and so on. The United States (and most of the world) has continued to use this single-issue approach, but in the 1980s, the Netherlands and a few other nations began to realize that it was not working.

The Dutch found that the single-issue approach might generate a very good air quality policy and still end up with a very bad waste policy; basically, it just shuffles problems around in the environment. A better future, they saw, would lie in trying to cross the barrier of single-issue approaches to create an integrated approach.

The Netherlands' earliest attempts at long-term planning were the "Indicative Multi-Year Programmes," begun in the early 1980s. There were about ten altogether, dealing with such issues as noise, soil, air, water, and waste over a policy lifespan of about five years. These plans served as an agenda for what ought to happen in the near future to change the quality of life in each particular medium.

But even as these programs were established, the Dutch realized that a more comprehensive, unified approach was required. One important catalyst was the United Nations' World Commission on Environment and Development and its 1987 report *Our Common Future*, better known as the Brundtland Report. This report has had a major effect on the environmental thinking of most western nations, and particularly affected the Dutch.

In many ways, the final Netherlandic plan was a response to the Brundtland Report and its concept of sustainable development – development that does not harm the environment's carrying capacity, or ability to renew itself. As we have seen, however, the term sustainable development does not really translate into practical, implementable policy. The Dutch have found sustainable development more useful as a way of helping them set goals for environmental recovery.

They also use the Brundtland Commission's concept of solidarity between generations, of each generation's responsibility to manage the environment in a way that sustains it for the next generation. A key psychological theme of the Netherlandic plan is that each generation cleans up. This concept of sustainability across generations is at the core of the NEPP, along with two other crucial components: a comprehensive, cross-boundary approach, and the integration of environmental issues into other fields of policy making.

One of the tools the Dutch used to develop their integrated approach was a century-old law called the Nuisance Act. This law was originally designed to deal with the disturbances, like noise or odors, that any human activity might create in a community. However, it did not consider the ecological impact of noise and odors.

In the late 1980s, the Dutch began to transform that act into the centerpiece of their integrated approach. In its new form it says that a single facility should be considered in terms of its total impact on the environment – in other words, it requires consideration of a facility's impact in the context of the goals for clean air, clean water, clean soil, and so on, before a permit is drafted for that facility.

The NEPP was the eventual outcome of the move toward integration. It was presented to Parliament in its final form by the Lubbers government in 1989. It contained 223 policy changes aimed at reducing pollution and establishing an economy based on sustainable practices. It was designed as a single, coherent, comprehensive policy that integrates all environmental areas and is based on an ecosystem approach. With the NEPP, the Dutch no longer simply react to a single incident, but instead consider what goals their society as a whole should be trying to attain.

NEPP policies cut across governmental and social lines and deal with all environment-related policy fields – economic, energy, agricultural, physical planning, and so on. The NEPP is also comprehensive in that it deals with all sectors of society, based on the premise that all of society, not just the government, is responsible for cleaning up the environment and preventing pollution. Because every single actor contributes to the problem, each is a stakeholder in the recovery process and is equally responsible for the solutions. This psychological approach is fairly new to the field of environmental policy making, and encourages more complex and sophisticated solutions to environmental problems.

In addition to the national plan, the government has required each of the twelve provinces to write a plan for its jurisdiction, with an update every four years. Provinces in the Netherlands are not sovereign, as are Canada's, but they do have regional jurisdiction, and they will, for the most part, be responsible for the implementation and enforcement of environmental policies.

Because comprehensive environmental planning is a process, not a one-time effort, the NEPP is updated every four years. The first update was the National Environmental Policy Plan Plus (NEPP+), covering 1990 to 1994. Early in 1994, the second version of the NEPP was published. While the first NEPP set out the main objectives of Dutch environmental policy, NEPP 2 focuses on carrying out the first plan's initiatives and ensuring that its goals are met. Core elements of NEPP 2 include strengthening the implementation framework, introducing follow-up measures where targets are not being met, and working toward sustainable patterns of production and consumption. International diplomacy and issues of economic development play a larger role in the later NEPP. In March of 1994, parliament reviewed NEPP 2 and ratified it with overwhelming support from both parties and from all sectors of the public. This was a very important step in securing the future of the NEPP in Dutch society.

When an inventor designs a new technology, he or she generally does so with the intent that it will actually work, and work well; the Dutch have designed their environmental plan with the same intent. The planning process in many of the developed nations has become an excuse for not making a political decision, or for putting off tough decisions. But when a government makes 223 genuine policy changes, and the voters back it up, the nation has made a serious commitment to change.

Making Sustainability Workable

The Netherlands' plan outlines these three particularly important mechanisms for making sustainability achievable: integrated lifecycle management, energy conservation, and improved product quality. General goals in regard to these, such as doubling the lifespan of products and stabilizing the use of energy by the year 2000, are stated in the plan.

Integrated lifecycle management goes beyond simple recycling or reuse programs; it is really about closing economic cycles and resource-use cycles in a way that makes them sustainable. There are many graphs and flow charts and illustrations in the plan about cycles and leaks and contributions, but what it all boils down to is reducing the quantity of materials used by industry across the board.

Energy conservation is also essential to sustainability, since developed nations are using energy at a rate that will use up current supplies within a few generations. The Dutch have set aside about $385 million per year for energy conservation over the next four years.[4] The national energy policy also requires an energy efficiency improvement rate of at least 2 percent per year,[5] and the installation of wind turbines with a generating capacity of 1,000 megawatts by the year 2000.[6] Ideas like cogeneration, using the energy from production for heating purposes, are very popular in the Netherlands, although so far they have only been implemented on a small-scale basis.

The third mechanism is improving product quality, so that products themselves have a longer lifetime and are easier to recycle or dispose of when they are no longer usable. For example, the Dutch government now requires the auto industry to take responsibility for what happens to cars when their useful lives are over. Among other things, this has forced the manufacturers to figure out what can be done at the drawing-board stage to ensure that it is easier to dispose of an old car.

It is challenges like these that bring out innovative thinking. For instance, Volkswagen made one small change that has helped a great deal: it marks all the different types of plastics it uses in the car, so that when it is demolished it is easy to tell what type of plastic the bumper is, and sort it appropriately. This means that much more of the plastic in old Volkswagens can now be reused. Before, when the plastics were mixed, it was difficult and costly to reprocess them.

The challenges have also led to the development of innovative planning tools. For instance, the Dutch government has established a fifty-year program for developing sustainable technology that does something called "backcasting." Backcasting asks what needs to be done today and in the next forty to fifty years to create a sustainable economy and technology by the year 2040. Backcasting means studying such questions as where it makes the most sense for the government to spend money – what sorts of new materials it should be trying to develop, or which kinds of alternative energy sources. Backcasting is geared toward developing the resources, the knowledge, the technology, and the products and services that will need to be in place by the year 2040 if the country's technology infrastructure and base is to be sustainable. This is no easy task. They are starting with a couple of demonstration projects to see if backcasting actually works, and will take it from there.

Important Principles of the Dutch Plan

There are a number of fundamental principles the Dutch have adopted in order to achieve sustainability, and which are incorporated into the NEPP. These are:

1. the stand-still principle (environmental quality may not deteriorate)
2. abatement at the source (remove causes rather than ameliorate effects)
3. the polluter pays principle (the user of a resource pays for the negative effects of that use)
4. prevention of unnecessary pollution
5. the application of the best practicable means (following the development of abatement technology)
6. carefully controlled waste disposal
7. application of a two-track policy of more stringent source-oriented measures based on effect-oriented quality standards
8. internalization of environmental concerns (motivating people to responsible behavior)[7]

Some of these principles are very well-known and widely accepted. Planners in our own country and in other developed nations agree that they are important principles, but rarely, if ever, apply them – and certainly their efforts are not comparable to what the Dutch have done by integrating all eight principles into one plan. Every instrument of the plan is required to take each of these principles into consideration, which is very tough to carry out politically and is the reason it has not been done before. The Netherlands' willingness to face such difficult political decisions shows that it and other green plan nations are for the first time becoming as serious about surviving as the superpowers have been about the development of technologies for destruction.

Although the principles listed may seem simple in theory, their application is extremely complex. For example, everyone would probably agree that unnecessary pollution should be prevented, but what are the "best practical means" for achieving that? In addition, the government's definition of unnecessary pollution will almost certainly be different from industry's, and each will propose a different way of dealing with it.

Translating these principles into action can mean major social changes.

For instance, the Dutch have taken principle 6, "carefully controlled waste disposal," and created a policy that no waste shall be exported; whatever is produced in the Netherlands must be taken care of there. This is a renunciation of the practice by some countries of shipping their trash to poor nations and dumping it for a fee.

Principle 7, the two-track approach, using both effect-oriented and source-oriented techniques, is primarily about flexibility. According to this principle, some problems are better handled from an effect-oriented perspective, which means applying cleanup technology, and some are better handled by source-oriented means, which focus on pollution prevention.

The Dutch are currently shifting their efforts toward source orientation, which is much more efficient in the long run, but they still use many effect-oriented measures, especially in terms of safeguarding environmental quality standards in a specific area. Any plan or policy must comply with the water quality standards of the water board, or with the air quality standards, which are national, or with the soil protection criteria.

Another key principle is that of internalization, which means that people must consider beforehand whether or not any decision they make or action they take will be consistent with environmental policy. It is probably the hardest to apply – asking CEOs or plant managers to take into consideration all of the principles listed above in the way they deal with all the environmental problems on their plates. And the Dutch do not just demand this of business; the same is required of government officials and their departments.

Planning by Scale and Theme

The Dutch use a five-level model of geographic scale to provide a framework for managing environmental problems. This model reflects the fact that every environmental problem originates on a particular geographic scale, and each solution is also located on a particular geographic level (although they are not always the same). The five scales they have defined, which come from the scientific report *Concern for Tomorrow*, are local, regional, fluvial (watershed), continental, and global.

The NEPP also sets targets and goals for specific policy issues, or themes, that are to be addressed at each geographic level. The themes also come from *Concern for Tomorrow*, which said they were the crucial issues for the future. They are:

- climate change (the greenhouse effect, damage to the ozone layer)
- acidification
- eutrophication
- toxic and hazardous substance pollution
- waste disposal (solid waste, radioactive waste, sewerage, soil clean-up)
- nuisance (noise, odor, safety)
- dehydration (water depletion and the draining of wetlands)
- squandering (depletion of soil and other resources)

An example of a problem on the global level is the issue of climate change. Although the Netherlands would be greatly affected by global warming, with 27 percent of its countryside below sea level and almost 50 percent less than one meter above sea level,[8] it cannot solve the problem of climate change itself. If other nations do not follow its example, the Netherlands is doomed. That is why it is so heavily involved in environmental education and negotiation at the diplomatic level.

Such problems as transboundary pollution from air toxics are continental in scale; acidification is one example. The NEPP's strategies to combat acidification focus on enforcing stringent emissions standards for all the sources that contribute to acid rain, such as auto exhaust, coal-burning plants, and so on. Other strategies promote reductions in neighboring countries, because most of the airborne toxics in the Netherlands come from somewhere else, just as many of its emissions end up somewhere else. If the Netherlands reduces its emissions it does not help its own cleanup that much, although it does help Norway's air quality.

Almost every country has problems with airborne toxics that drift in from other places, so international cooperation on this and other environmental issues is crucial. That is why there are treaties like that of the UN Economic Commission for Europe on transboundary pollution and acidifying compounds. This treaty covers all of Europe, dealing with these issues and seeing that each country does its share, and in the end everybody contributes and everybody benefits.

Fluvial problems are on the watershed level; deforestation is one example. The Netherlands does not really have fluvial problems apart from the Rhine River basin, but that is probably one of the biggest issues it faces (see fig. 3). The quality of the Rhine water, which accounts for more than 50 percent of the country's drinking water, is sometimes so poor that the intake

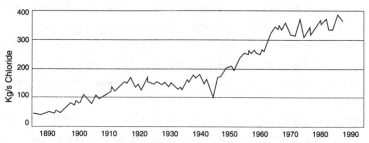

3. Chloride Load of the Rhine at the German-Dutch Border, 1885-1985. The Netherlands suffers from the emissions of neighboring countries, as well as its own. This graph shows the increasing levels of chloride contamination in the Rhine at the German-Dutch border over the last century. Source: The Netherlands' Ministry of Housing, Physical Planning and the Environment, *Essential Environmental Information, 1991.*

pumps must be shut off.[9]

The problem of eutrophication operates on both the fluvial and regional levels. Eutrophication, which is the depletion of oxygen in water sources due to the accumulation of large amounts of nutrients, is usually caused by agricultural runoff, and is a big problem in the intensively farmed Netherlands. Most Netherlandic farmers do not have enough land to dispose of all the manure produced by their animals; it must be treated or processed into products. Part of the NEPP strategy for this theme is to get farmers to cut back on the amount of fertilizer they use, but in the end, they may have to scale back their numbers of cattle.

The diffusion of toxic substances in water and soil, another problem that operates on both the fluvial and regional level, is one area in which the Dutch have not yet made a great deal of progress, primarily because their pesticide reduction programs have not been fully implemented. Because farming in the Netherlands is so intensive, there has not been much support for switching to organic farming methods. The government is concentrating now on switching over farmers to less hazardous alternatives, and convincing them to use smaller quantities of pesticides.

Much of the country's soil is a type of clay that is quite productive. Unfortunately, clay also holds toxic compounds longer than do sandy soils, and that creates another set of problems. It means that the active compounds in pesticides will trickle down to the groundwater level and eventually to the drinking water – and some have a lifetime of twenty to thirty years. The Netherlands is encountering drinking water problems now from

57

substances used decades ago, and that is a problem in itself, because it means that in some areas of the country people are living on a time bomb.

Then there are local problems, like odor and noise, which are important issues in the Netherlands because its population is so dense. Local problems also include the indoor environment – things like asbestos, formaldehyde, and secondhand smoke.

The NEPP sets specific targets for pollution reductions for each theme and at each geographic level. For instance, on the global level it sets a goal of stabilizing the Netherlands' carbon dioxide emissions at 1990 levels by the year 2000 (this has since been changed to 1995). On the fluvial level it aims for a 70 to 90 percent reduction in emissions of eutrophying or poorly degradable substances by 2010, while on the regional level it calls for a 70 to 90 percent reduction of eutrophying, acidifying, and nondegradable substances. The size of the waste stream is also to be reduced 70 to 90 percent.[10]

If each theme, or problem area, has an appropriate level of scale, so do each of the actions to be taken in response. For example, a problem could have global effects, but if there is only one source – an accident at a nuclear reactor, for instance – then action should be taken at the local level. Rather than treating people for radioactive fallout, it would make far more sense to go to the source of the problem and stop the emissions from the reactor.

The instruments chosen should also be more or less determined by the scale of the problem as well as of the solution. For example, there is no point in putting money into demonstration projects on climate change; it makes more sense to come up with treaties that require every country to cut its emissions a certain amount within a set period of time.

The Target Groups

In creating the NEPP, the government wanted to tailor its policies to those social sectors that contributed the most to the problems listed above. It therefore selected a set of target groups, most of which are composed of industries in various sectors, that were very carefully chosen on the basis of the contributions they make to one or more of the themes.

The key target groups are: agriculture, traffic and transport, industry, the energy sector, the building trade, consumers and the retail trade, the environmental trade (water suppliers, the waste sector), research and education, and public organizations. Some of the business-related target groups have been broken down even further, into specific business categories such as the basic metals industry, the packaging industry, and the printing industry. The government has set both general and specific goals for these groups.

Much of the Netherlands' environmental policy is set by means of signed agreements reached between the government and representatives of the various target groups. Often the industry representatives are allowed to determine the most efficient way to reach the goals set by the scientists, rather than the government imposing rules that might or might not work.

The long-term goals for each target group are not negotiable, but the methods for getting there are. The long-term perspective is important in this process of cooperation, because it gives industry time to adjust. In most cases an industry cannot just change overnight; this would cause a great deal of disruption and difficulty, and might not even be possible. But if you tell it where it needs to be in the year 2010, there is then room for negotiation and compromise regarding the short-term actions needed to accomplish those goals. That is a very different, very important element of the Netherlandic approach, and it has been so successful that just about every goal in "To Choose or to Lose" has the blessing of the industries concerned – although there is plenty of fighting over the short-term methods.

The government has encouraged the companies within each group to organize their efforts in a trade association, to make negotiating and policy making easier and more efficient. The government also realized that if it wanted industry's active cooperation toward environmental goals, it would have to make it easier for businesses to work with government. To do this, it appointed a target-group manager for each of the target groups, an official in the Ministry of the Environment to whom businesses can bring all their questions and problems regarding environmental issues. The efficiency of this "one mailbox" approach is common to the green plan nations. It secures businesses' initial cooperation, and convinces them to compromise on a wider range of issues.

The government's ability to secure business's cooperation is a remarkable accomplishment, and an essential part of the NEPP. That is why I believe such plans will work in the United States and other developed nations, where business is frustrated by current environmental regulations and laws. Furthermore, there is no lack of distrust between industry and government, and between industry and environmental groups. Green plans offer a workable alternative to this old dynamic, which has been in no one's best interest. The biggest factor behind the success of the existing green plans has been the willingness of businesses to sit down and negotiate environmental quality goals and regulations, and their willingness to cooperate in putting together a plan.

In return, nations that adopt these comprehensive environmental strategies will need to help their industries make the changes and maintain their competitiveness. On the microeconomic level, the Dutch do expect shifts in some industries, with some companies going under for the simple reason that their business is so destructive to the environment that the country no longer wants it. So far, however, no business the government knows of has moved or gone bankrupt because of environmental policies.

One of the great advantages to this type of planning approach is that it ensures a relatively stable regulatory environment. The Netherlands' environmental policy may be tough, but it does not spring any sudden surprises on business. Another competitive advantage the Netherlandic approach offers to business is in the field of environmental technology. In many cases, their companies are on the cutting edge of environmentally sound products and practices, because of the NEPP's strict requirements. The international market for these technologies is growing rapidly.

The target-group approach to business is a major innovation in environmental policy setting. Only its main elements have been summarized in this section; because it epitomizes the new type of government-business partnerships that green plans promise, it will be discussed in greater depth in the chapter devoted to that subject, chapter 9.

Nonindustrial Target Groups

Not all of the target groups are entirely composed of businesses; some include individuals in their roles as consumers, while others include educational and research institutions, or environmental organizations. These nonindustrial target groups pose their own challenges for the government in terms of developing strong, realistic policies.

For instance, trying to reach consumers as a group, in the way they consume products and services that have a specific impact on the environment, is very different from communicating with other target groups. It is not possible to sit down with individual consumers to negotiate a relationship, as can be done with industry, and consumer groups do not represent consumers in the same way a trade association represents its member companies.

One way the Netherlandic government has reached out to consumers is through a massive public information campaign. Although average citizens have long been aware that the Netherlands has serious environmental problems, they did not know specifically *what* was wrong, nor what the sources of the problems were. They certainly did not realize that they themselves were contributing to the problem, in the way that they used energy and water, or did not separate their waste.

To correct this, several years ago the government established a public information program, using magazines, television, radio, billboards, and leaflets in the public libraries and post offices. The campaign was in three parts: the first focused on building more specific awareness; the second indicated specifically what the problems were; and the third concentrated on the actions each individual could take to correct the problems.

This last element of the program focuses on changing the day-to-day behavior of the average citizen to a much greater extent than has been attempted before. First, they are really trying to instill a commitment to stewardship; and second, they are giving people the information and tools needed to act on that commitment (chapter 10 provides a more in-depth discussion of this idea of changing behavior).

To provide more information to the consumer, the government has adopted programs such as eco-labeling for environmentally friendly products and logos that identify how to dispose of certain products. The Environmental Advertising Code came into force in 1991, forbidding misleading claims about the environmental friendliness of products.

The Dutch have another target group made up of societal groups, primarily environmental nongovernmental organizations (NGOs). Labor unions and employer associations are also included. In the Netherlands, as well as in a few other European countries and in Canada, some of these organizations are partially funded by the government to perform their roles as advisors and critics. They play their roles as critics quite well, even though they get some government money; these groups have no problem suing the government when they think it is necessary. Of course, the relationship is not just adversarial; NGOs also provide advice and education. I met some representatives of Netherlandic environmental groups at the Rio Earth Summit, and they were very proud of their country's efforts.

Research institutions and educational organizations have also been identified as a target group, and asked to adopt programs that reflect the intensified need for environmental research and policy making.

Monitoring and Updating

One of the NEPP's management strengths is its built-in mechanisms for careful monitoring of progress and setbacks, and for the frequent updating of goals and strategies based on this information. The basis of the long-term effort is the NEPP plan itself, a 400-page document detailing all the myriad policy changes, which is to be updated every four years. These four-year updates are forecasting reports, focusing on projections of what will happen down the road. The reports study such questions as whether or not the plan is still on track – whether or not the methods they have adopted so far are going to be sufficient. They ask such questions as: Should new themes be added? Are there new problems emerging that need to be addressed?

In the second year of the four-year period a report card on the plan is published. These report cards are more reflective, looking at what is actually happening and at what the implications are for future policy. The first report card came out in 1992,[11] and is a remarkable document, 533 pages long. It looks at such questions as how the plan is being enforced and how the different measures are contributing to the ultimate goal of solving the problem.

These reports are the outside control on what the government is doing. Although the government pays for the reports, the research institute that actually does the reporting – the same one that wrote *Concern for Tomorrow* – is independent and highly respected. Regular outside audits of its programs give a government a strong management advantage, because every time a manager or official signs a document, he or she knows that it is going to be held up to daylight sometime within the next year, when the reviews are done. Results tend to improve when the people in charge know that they will be held accountable for their decisions.

In addition, the government has its own yearly reports, called environmental programmes. These provide the short-term oversight on what the government has been doing, and what the agendas for the next year or two years will be.

The honesty of the Netherlands' monitoring and reporting is remarkable. The first four-year NEPP update clearly reflects the complexity of the situation and the importance of the plan's comprehensiveness. The government manages to deal with the entire package, functioning at a level of thinking that we in the United States have not even begun to comprehend. The update exposes the scientific truth of what is really happening under the program, both good and bad, and there have been plenty of examples of both. Honesty and commitment is part of the reason the public is so strongly behind the plan, why it understands and has confidence in it.

Achievements, Setbacks, and the Next NEPP

The Netherlands has achieved some remarkable results with its comprehensive environmental policies over the last decade, and implementation of the NEPP promises more and faster progress toward environmental recovery. Some of the most easily quantifiable gains have been made in the area of emissions reductions. For example, during the period from 1980 to 1990, emissions of the main acidifying substances, sulfur dioxide, nitrogen oxides, ammonia, and volatile organic compounds, were reduced by 30 percent, and phosphorus and nitrogen emissions from wastewater and agricultural runoff decreased by 5 percent.[12] As of 1992, pesticide use had been reduced by 20 percent from 1988 levels, and, at the time of this writing, chlorofluorocarbon (CFC) use has been virtually eliminated in the Netherlands (see fig. 4).[13] By the end of 1995, discharges of industrial wastewater into the Rhine are expected to be 50 to 70 percent less than in 1985, and by the year 2000, emissions of volatile organic hydrocarbons are expected to fall by 60 percent from 1990 levels.[14] Although all the NEPP goals will not be met, the upward trend for most emissions has been reversed.

The comprehensive approach has also allowed the Dutch to create some innovative new programs. One that has caught the world's attention is the flooding of a part of the farmland they drained a century ago.[15] Previously known for building dikes rather than breaching them, the Dutch plan to eventually restore about 600,000 acres of wetlands, with the dual goal of returning the land to the wildlife that once lived there, and also of halting the fall of their water table.

They also have an ambitious plan to plant almost 99,000 acres of forest, as part of their objective to eventually be self-sufficient in timber and to make a contribution to the reduction of carbon dioxide in the atmosphere. Wildlife habitat will be expanded by almost 50,000 acres. These kinds of large-scale projects would be next to impossible without a comprehensive and integrated

Figure 4

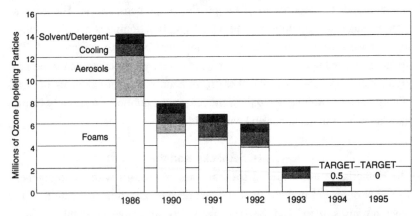

4. Time Scale for the Reduction of the Use of CFCs in the Netherlands. The NEPP has allowed the Dutch to make rapid progress in some areas. As of this writing, CFC use had been virtually eliminated in the Netherlands, and pesticide use was reduced 20 percent between 1988 and 1992. Source: The Netherlands' National Institute of Public Health and Environmental Protection.

plan, because they involve several ministries.

Cleaning contaminated soil is another fresh approach the Dutch have adopted. There are serious dangers to environmental health from contaminated soil, where the pollutants have exceeded what is safe for growing crops, or in cases where the toxics leach into the groundwater. One of the first initiatives the government undertook in partnership with business was to list all sites where the soil had been contaminated and develop cleanup plans. And although it is a costly process, it will have to be handled as part of their overall budget planning. In the past, the government has funded most of the soil treatment, but in the coming years polluters and users will contribute more to the costs.

For the period 1991 to 1994, the Dutch estimate that more than 10 million tons of soil will have been treated. They estimate that between the years 1991 and 2010, the total amount they will need to spend for soil treatment will be a minimum of $7 billion (the cost will be shared among government, industry and users).[16] It is encouraging to realize that, while it is not possible to create new soil, it may be possible to clean up soil that has been contaminated. We should never forget that a civilization depends on the vitality of its soil.

Another example of the government's innovative thinking is its agreement with the car manufacturers of Europe, which required the manufacturers to take responsibility for making sure that their cars are disposed of in ways that will have the least impact on the environment. This is a lifecycle management approach and, as mentioned above, has led to such innovations as Volkswagen's clear marking of plastic car parts for recycling.

But this policy cuts even deeper, by saying that it is Volkswagen's responsibility to ensure that an old Volkswagen be dismantled in the proper way. This does not mean that Volkswagen will now be dismantling cars; instead, Netherlandic car importers (the Netherlands has no auto manufacturing industry) have negotiated a deal with the businesses that dismantle cars and sell parts. These facilities will dismantle all types of cars, and will have to be certified by the government. In the future, Volkswagen will be able to show the government that it is taking care of its responsibility because its cars are being dismantled at a government-approved, environmentally sound facility.

The NEPP's target-group approach has in particular yielded some unexpected and very encouraging benefits. For example, the energy distribution target group produced an environmental action plan that achieved 90 percent of its carbon dioxide reduction target in its first year, and subsequently set itself even more ambitious goals.[17] A number of other industries have also chosen to set themselves more ambitious targets than those required by the NEPP.

In some cases, the target-group approach has led to greater cooperation among industries in order to reach environmental goals. One example is the packaging industry, where manufacturers, importers, distributors, and retailers have joined forces to increase their efficiency in achieving the targets required of them.

The Dutch have also encountered setbacks, of course. They were not able to reach their initial goals for carbon dioxide reductions because the greatly reduced price of energy in recent years has encouraged the number of cars to increase at a greater rate than they had anticipated, and has also allowed people to keep driving less fuel-efficient cars. They are considering imposing a carbon tax, but prefer to do so in conjunction with other members of the European Union.

Another example of the kind of initial failures that can be expected in a plan of this scale is their inability to process enough cattle manure to meet the goals they had set. To be on target, they would have to process at least 5.5 million tons by 1995, whereas they will at best be able to handle only three million tons by the end of 1994.[18] But this also demonstrates the importance of computer-based information; they are able to check their progress and see where they are falling behind as the program proceeds. An additional setback is the Netherlands' sudden rise in population, due to immigration from developing countries and from eastern Europe. That growth has made it more difficult for them to achieve some of their goals.

Despite these difficulties, the latest reports the government has received show that NEPP implementation is on schedule and the targets set in most issue areas will be met. Those that will not be met will be addressed by NEPP 2, once that is fully implemented.

The government learned a great deal from the successes and failures of the first NEPP, and that experience has been incorporated into the new plan. Some of the strategies developed for NEPP 2 include:

- expanding the target-group approach to other sectors, and creating clear targets and tasks for each sector
- using a broader mix of instruments and selecting the appropriate instrument for each situation
- improving the flow of environmental information
- providing the education, infrastructure, and facilities necessary for good environmental choices (for example, providing convenient recycling points)
- providing clear criteria for determining environmental priorities
- strengthening regional policy approaches through the development of integrated regional plans
- achieving a better integration of enforcement issues into the policy-making process
- pursuing active environmental diplomacy, particularly at the level of the European Union
- promoting debate on the social and economic implications of sustainable development, and developing the concept of eco-space
- encouraging further research, development, and demonstration of new technologies[19]

Probably the most distinctive feature of the NEPP is its technical superiority, which has as its foundation a very good information base. The decision making, and the advanced planning approach, are based on that centralized, readily available information. The Netherlands needs this kind of approach in order to manage the intense concentration of people and activities. In contrast, Canada and New Zealand, which are much more open landscapes without the dilemma of population density, have not had the need to develop their information bases so highly.

Much of the information is collected in a Central Bureau of Statistics. The government also has a standing agreement with a consulting firm to constantly update a database on the cost of environmental technology. One of the things the Dutch have been able to do with this information is define "sustainable level points" for a number of their themes. These are points on a graph showing the sustainable level of pollution for a particular theme. Anything above that point is unsustainable, so the goal is to stay below the line in all the themes. They do not yet have the scientific knowledge to define sustainable levels for every theme, but they were able to calculate some of them for the first time in their 1992 environmental report.[20]

Another distinctive element of the Netherlands' approach is their honesty in reporting. I have never seen a government stand up and say so clearly, "We failed at this." They are brave enough to actually say what they are doing, in a way that every citizen can easily understand – to show a line on a graph that should go down and does not, or is not going down as fast as you would like. They are very open and honest, and that takes a great deal of political courage.

• • •

In Amsterdam's Rijksmuseum, known for its collection of paintings by the Dutch masters, there is one painting that looms above all. It is by Rembrandt, and it forever changed the world's ideas about art. It is titled "The Night Watch," and it is painted in three dimensions; when you look at it, it appears that the hand is actually reaching out to you. Rembrandt startled the world with that painting, so much so that it changed our ideas of art forever.

I believe that the NEPP will similarly change the way people see themselves and their actions in relation to the environment. And because I care so much about finding an answer to environmental decline, I believe the individuals who have come together to form the Netherlands' plan should be remembered as having created a work at least as important as Rembrandt's.

5

New Zealand Starts from Scratch

New Zealand recently enacted a green plan, called the Resource Management Act (RMA), that makes it one of the world's leaders in environmental policy. The RMA emphasizes the two key themes of green plans, comprehensiveness and integration.

The RMA is a remarkable achievement, primarily for three reasons. First, it is based on the maintenance of a high level of quality of life for the residents of New Zealand. Second, it made major, almost revolutionary, changes in government, entirely restructuring New Zealand's resource agencies and laws around the premise of what they term "sustainable management." A prime example is that resource management in New Zealand is now based on districts defined by watersheds, rather than arbitrary political boundaries. Third, it raised the level of public awareness and involvement in the process of reform. The changes entailed by the RMA would not have been possible without substantial public support.

Like the Netherlands, New Zealand is shifting away from regulating exactly how resources are to be used, toward regulation of the effects that resource use will be allowed to have on the environment. But the Dutch, because of the state of their environment, are focusing primarily on the recovery levels they want to achieve in the next twenty-five years; they have done less to develop an underlying philosophy of how and when economic activity is to be accommodated. The RMA is unique among environmental policies and plans in that it does just that. With the RMA and its philosophy of sustainable management, New Zealand has developed a carefully thought-out framework for structuring all its present and future decisions about resource management – what its values, goals, and considerations should be.

The momentum for New Zealand's plan came in part from the public's reaction to an earlier administration's push for massive development in the late 1970s and early 1980s. Part of this proposed development was a series of energy-related projects, including a synthetic gas-to-gasoline plant, a methanol plant, and a major dam on one of the country's wild and scenic rivers. In order for these projects to be built, the government proposed changing a number of environmental laws, including limiting the right of individuals to testify.

This "Wellington knows best" attitude (Wellington is the capital of New Zealand) touched off a public furor that eventually brought about a change in government. The prime minister who advocated the development, and his party, were replaced by a more environmentally oriented party, which came into office intent upon changing and upgrading New Zealand's environmental laws.

The new deputy prime minister, and later prime minister, Geoffrey Palmer, also became minister of the environment. This dual authority, and a special committee he established comprising the key ministers of finance, local government, commerce, energy, and transport, gave the government considerably more power to make major environmental reforms. It was this government that initiated the RMA . The process took a great deal of time, though, and despite bipartisan support for the RMA in Parliament, Palmer and his colleagues were unable to pass it into law before the government changed again.

The new government was formed by the more conservative party, which was initially concerned that the reforms did not represent an adequate balance between the environment and the economy. But the public was solidly behind the reform process, so the new government took up the legislation and even made improvements in it.

When the RMA was reintroduced in Parliament, it passed with flying colors, supported by both liberal and conservative politicians as well as a broad cross section of the public. It will be difficult for a new government to stop the program, even if it wanted to. The law itself may – and probably should – change over the years, as new problems are faced and new information is gathered, but the overall purpose of recovering environmental quality is firmly planted and growing stronger in New Zealand society.

New Zealand's Environment

New Zealand is an island nation, about the size of the state of California, in the South Pacific Ocean.[1] Many rare plants and animals evolved there, in isolation from the larger land masses, including many flightless birds. When the first humans to arrive in New Zealand, the Maori, landed about 1,000 years ago, approximately 80 percent of the country was forested.

Although the Maori initially burned some areas of the country and caused some bird extinctions, about two-thirds of the original forest cover remained when the Europeans arrived about 150 years ago; since that time, much of this has been cut or burned. The Europeans also turned much of the land to pasture or farming, built towns and roads, and introduced new plants and animals, all of which had significant impacts on the country's ecosystems.

New Zealand's current major industries – farming, forestry, horticulture, fishing, minerals extraction, and, more recently, tourism – are highly resource dependent. However, most of its population is based in urban centers. This means that, while the economy is largely rural-based, the population's attitudes and lifestyles are primarily urban.

Most of the country's electricity, about 70 percent, comes from hydroelectric power; the rest is from gas and some coal. Fuel for vehicles is largely imported, although some is produced from natural gas. New Zealand also has an alternative fuel industry that powers about 15 to 20 percent of the vehicle fleet with propane or methane.

Fortunately, most of New Zealand's environmental problems have not yet reached the level of many northern hemisphere countries. This is due in part to its smaller population, and also to its relative lack of heavy industry. New Zealand has relied on its resource-based economy rather than on heavy industry, trading for the industrial products it needs.

However, this nonindustrial economic base has led to environmental problems caused by the pressures put on resources. Instead of suffering from waste from mismanaged steel plants, for example, New Zealand's environment has been damaged by poor practices in the farming and forestry industries, and from the introduction of nonindigenous species into its ecosystems.

As in other developed nations, unsustainable agriculture has been a problem for New Zealand. Modern agricultural practices such as monoculture and the heavy use of fertilizers, pesticides, machinery, and irrigation have all had a negative impact on the country's soil, water, species diversity, and natural ecosystems. Until recently, New Zealand's main industry was livestock farming, primarily sheep. Sheep monoculture can be very hard on farmland, particularly with New Zealand's emphasis on high levels of production, and especially on marginal farmland with a high risk of erosion. Despite the negative impacts it entailed, the practice of clearing marginal land in order to raise more sheep was once subsidized by the government. Much of this has now changed, due to the government's decision to end farm subsidies.

Some of New Zealand's most serious problems involve threats to soil, including erosion, loss of fertility, compaction, pollution, flooding, and urban encroachment. About 50 percent of the country's land area shows signs of erosion; of this, approximately half is slightly eroded, the other half moderately to severely eroded. Some of the severe erosion in mountainous regions is due to natural causes.

Wildlife is an especially significant indicator of New Zealand's environmental quality. Because many primitive life forms have survived there, New Zealand's indigenous plants and animals are of international importance. But many of these native species are threatened or endangered – some 18 percent of vertebrates and 16 percent of native flowering plants – and the extinction rate is reportedly among the world's highest.[2]

Because of inadequate regulatory systems, some types of fish and shellfish have been overexploited, and the related fisheries are declining. Development has devastated wetlands, and drainage, reclamation, and pollution have taken their toll as well. Waste management practices are another problem area for New Zealand. With a small population, volumes of waste have been low, and landfills sites have been chosen for convenience rather than for ecological reasons. It is becoming difficult to find suitable new sites in and around the major urban areas.

The Political Basis of the RMA

The actual process of changing New Zealand's environmental laws began in 1988. Both government and citizens realized that they had been passing laws for a hundred years or more that were overlapping, contradictory, unclear, or riddled with gaps. The laws simply were not accomplishing the purpose of protecting the country's natural resources.

The government realized it was pointless to try to reform this tangle of regulations and interventions piece by piece, and decided to start again from

1840

■ Indigenous Forest

100   0   100   200   300
Kilometres

Map 1. Estimated Indigenous Forest Cover of New Zealand, 1840 and 1976. European settlers
have cut or burned a significant portion of New Zealand's indigenous forest cover since their arrival
about 150 years ago. Many of the country's threatened species depend upon the remaining forests.
Adapted from Wendelken, 1976 Forests, New Zealand Atlas (ed) 1. Wards. Reproduced by permission of the Department of Survey and Land Information, New Zealand. Crown Copyright Reserved.

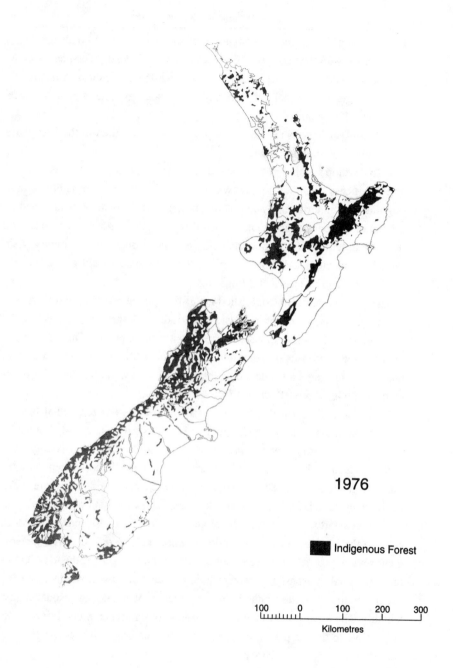

1976

Indigenous Forest

100   0   100   200   300
Kilometres

the beginning. It initiated a three-phase review process, to be conducted by the government with the participation of all segments of society. The phases were:

- to ask and answer the question of whether there should be a law for the management of natural and physical resources, and which resources should be included
- to look at options for achieving the objectives defined in the first phase, and choose the best means
- to develop specific policies and draft appropriate laws[3]

The process of public review was remarkably thorough, possibly as thorough as any undertaken by any democratic nation in history. It involved all sectors of the public, and took at least three years; the government team left no stone unturned. It held hundreds of meetings throughout the country, visiting all major cities. After the hearings, the field team would go back to the home office and discuss its findings.

The government also established a toll-free phone number, and there was considerable discussion on television and radio. Altogether, the government received more than five thousand responses from the public. The result was a broad-based consensus on goals and the best means for achieving them. From that consensus came a single, coherent, consistent law, the Resource Management Act, which took effect on 1 October 1991.

New Zealand's form of government facilitated its environmental law reform. It has a unicameral legislature; that is, it has only one house. The leader of the majority party becomes prime minister, and the cabinet of twenty or so ministers is also chosen from the majority party. There is no written constitution and no federal system of provinces. The executive can therefore make major changes to laws relatively quickly; this is how the RMA was created.

New Zealand's small, largely homogenous population also made it easier to achieve the kind of consensus needed to enact radical reforms. The government review process discussed above built public support for the RMA, creating a national consensus. Although public education is still needed, most of the population now understands the importance of managing resources sustainably. The government intends to continue providing as much information as possible to the people, so they will see what the program is accomplishing and what actions are necessary.

Several influential social groups in New Zealand played particularly crucial roles in the creation of its plan, and will play equally large roles in its implementation. One of these is the Maori, citizens of Polynesian heritage who were the country's first human inhabitants. Because the Maori were entirely dependent on natural resources, they found they had to adopt sustainable practices in order to survive. It is believed that their society became ecologically sustainable some time in the seventheenth century.[4]

The Maori have been granted rights as indigenous people that are different from the entitlements of other citizens, particularly in terms of resources. For this reason, the management structure of the RMA has a component that deals directly with the Maori as a distinct political constituency. They also have a well-organized tribal structure and their own representatives in the national Parliament, and it is the government's intention that tribal authorities be consulted regarding resource management issues.

Business groups also played an important role in enacting the RMA, and are playing an even larger role in its implementation. As in most countries, New Zealand's business community tends to be conservative, and was initially suspicious of the RMA. But from the outset the government included key trade associations of each industry, particularly the primary industries, in the planning process. Far from having the RMA imposed on them, these groups took part in the process of creating it.

In a recent conversation, the representative of a cross-sectoral association of forty resource-using companies took the most aggressive stance I have heard from a business leader anywhere. He told me that business interests had at first fought the reforms, but realized in the end that they had to make a total commitment; they had to work together to establish a major environmental statement for the nation and really make a difference for the future.

They came to realize, he explained, that their old behavior of approaching government from a self-interested perspective had not worked. One timber company would manage to change the rules to its benefit, then a trucking company would come along and have the rules changed again. This manipulation to benefit narrow interests went on to some degree even when companies banded together in trade associations because they were still operating from a completely self-interested position. Seeing this, the companies in this association decided to do the idealistic thing and support the adoption of sound environmental practices, under which all businesses would be treated the same.

One reason that it was relatively easy for the government to get the leaders of the trade associations to agree to work with it on resource issues is because

New Zealand's economy depends so heavily on natural resources. It was particularly successful in enlisting farmers, who represent New Zealand's largest industry. The farm industry is now at the point where it sees the economic advantage in sustainable production, and in some cases it has gone beyond government requirements, setting higher standards of environmental quality.

A good example is the Federated Farmers, the farming industry's primary association, which has come to realize that environmental sustainability is not only the key to survival, but also a great marketing advantage. High-quality meat, dairy, and produce raised without large quantities of fertilizers and pesticides are attractive to many consumers in countries like the United States, particularly if they cost less than consumers are used to paying. New Zealand can produce these foods more cheaply than most countries because it has lower energy inputs for agriculture.

Not all industry leaders have accepted that eco-sustainability can be an advantage, of course, or even all farmers. But New Zealand has been fortunate enough to have a great deal of support from some of its sector leaders.

Nonprofit environmental groups also played an important role in drafting the RMA, participating in negotiations throughout the years it took to develop the Act. However, as befits the diverse nonprofit environmental movement today, some of the groups have remained skeptical, playing the role of critics. Compromises have to be made when creating and implementing a complex law like this, and it is important to have watchdogs who will keep up the pressure for improvements and alert the public to any defects.

Environmental groups also help the process along by public education. In New Zealand, the government has a program of environmental grants that gives funds to environmental groups to establish and maintain educational programs in communities across the country.

The Philosophic Basis of the RMA

The RMA represents a truly radical break from traditional approaches to environmental planning. At its core are two key philosophical differences. The first, which it shares with the plans of both the Netherlands and Canada, is the move from a narrow, piecemeal approach to a more comprehensive and integrated view of resource management. The second difference, which is unique to New Zealand, is the concept of sustainable management as the structuring principle behind the country's environmental laws and policies.

Like many other countries, New Zealand used to approach the management of resources as a technological challenge geared to solving one problem or one need, and would plan, for example, to build levees to stop floods or irrigation systems to supply arid regions. But, like other countries, it found that approach seriously flawed, often causing more problems than it solved.

For instance, they would deal with a flooding problem by looking at only one part of the picture – the area in which the river floods – and then building levees and dams to control it. But in the long run, that would be more expensive, less effective, and more environmentally damaging than if they had looked at the whole picture and realized that the answer was to plant trees on the steep, deforested slopes upstream, which would lead to much less water runoff and siltation. Or, taken one step further, if they had identified areas of flood hazard and provided that information to local authorities and the public, better decisions could have been made regarding the proper use for the land.

The RMA is designed to allow communities and regions to take that broader approach. It is a comprehensive framework that integrates the institutions and systems dealing with resources, so that the environment can be dealt with as a whole. Instead of dealing with the environment in an inefficient, piecemeal fashion, communities can now create long-term, comprehensive management programs for their resources, supported by the enabling legislation of the RMA, and by money and ideas from the government.

New Zealand's push toward comprehensiveness and integration led to a massive restructuring of government. The RMA replaced fifty-seven existing resource-related laws, and whittled eight hundred units of government (this included a multitude of little water boards and trust boards and so forth) down to ninety-three.[5] It consolidated government into two levels, national and local, then divided local government into regional authorities, which are based on watersheds, and district authorities. Both government and the regulatory process have been streamlined to make them more effective and efficient; the intent of the RMA is to get the maximum environmental benefit with a minimum of regulation.

As the New Zealanders discovered, these efficiencies could not be achieved without first determining the ultimate goal of their resource policy. During the review process, they identified this goal, the sustainable management of resources, and refined the concept through further debate. The definition they finally arrived at is stated in section 5(2) of the RMA as: managing the use, development, and protection of natural and physical resources in a way, or at a rate, which enables people and communities to provide for their social, economic, and cultural wellbeing and for their health and safety while:

(a) Sustaining the potential of natural and physical resources (excluding minerals) to meet the reasonably foreseeable needs of future generations;

(b) Safeguarding the life-supporting capacity of air, water, soil, and eco-systems; and

(c) Avoiding, remedying, or mitigating any adverse effects of activities on the environment.

All parts of the RMA, and any decisions made and actions taken under it, are required to meet these principles of sustainable management. But the act also recognizes that definitions of sustainability are not static and will change as the store of environmental information grows.

New Zealand's ultimate goal is sustainable development, but the RMA by itself does not attempt to provide that. Sustainable development, as the government's review group concluded, embraces a very wide range of issues, including social inequities, global redistribution of wealth, and population density problems. The group found this range too complex for a law designed to manage the natural resources of a single nation. Consequently, its definition of sustainable management does not address the question of development; it is neither anti- nor pro-development. In promoting sustainable management, the government is not as concerned with how the land is used as it is with how various land uses affect the environment and other people. They have shifted from planning for activities to regulating the effects of activities.

The RMA provides a structure within which decisions are made about the way community-owned and -managed resources are allocated, determining who is allowed to use such public resources as water, air, the coastal area, and geothermal energy. It also determines what they may be used for. But the act also applies the principle of sustainable management to privately owned property, setting the standards of environmental quality that private owners must adhere to when making decisions about the use of their own property.

This is a powerful change in thinking about environmental law and resource management, to apply quality control standards to every public and private decision about land use. Other countries do this to some extent, of course, through permits and regulations, but New Zealand is the first country to codify into law the idea that the government has a right to require private landowners to meet certain standards. The judicial system in the United States is just beginning to deal with this issue.

New Zealand's policy is not a strict code of behavior, however. Rather than telling people how to use their land through controls like zoning, the government instead gives people standards for environmental quality that must be met regardless of how the land is used. In taking this approach, the government is trying to achieve two almost contradictory goals: first, allowing people the maximum freedom in their use of resource; and second, ensuring that those activities have the minimum adverse effect on environmental sustainability.

Instead of saying, "You will put houses here" and, "You will put industry there," the RMA has set up a process for developing standards of environmental quality that will be consistent with sustainability, and which will be specified in fairly clear ways. Within that framework people have the freedom to do what they please.

How the RMA Works

Under the RMA, the national government has two complementary means for expressing and applying its resource management policies: national policy statements and national environmental standards.[6] Policy statements express national goals and objectives for the environment and its sustainable management; they are descriptive, rather than prescriptive, and cover issues of resource protection, use, and development. The statements may also deal with general issues, such as New Zealand's obligations in enhancing the global environment, or they may be quite specific about a particular issue or site.

The only national policy statement the government is required to develop is the New Zealand Coastal Policy Statement, dealing with national priorities for the management of the country's coastal environment.

Unlike policy statements, national environmental standards are prescriptive, and are promulgated as regulations. They apply to the entire nation; regional and local plans and policies cannot violate them. They set technical standards relating to the use, development, and protection of natural and physical resources, including standards for contaminants; water quality, level, or flow; air quality; and soil quality. Typically these will be bottom-line standards, beyond which one cannot go and still practice sustainable management, but they can go further in particular situations. National standards are still being developed at this stage.

Regional and district governments will bear the most responsibility for implementing the RMA. The act restructured regional government into sixteen units based primarily on watersheds and their ecosystems, which are the most logical unit upon which to base environmental management. Directly elected regional authorities are responsible for preparing regional policy statements and plans. All regional and district policies and plans must be consistent with and reflect national policy statements and standards, and all must uphold the principle of sustainable management.

Regional policy statements are mandatory, because they will articulate the key issues and priorities for each region, interpreting sustainable management and applying it to the region's biophysical and socioeconomic characteristics. They will identify key resources and their condition, determine the community's relationship to and dependency on those resources, and identify links among resources and ecosystem issues and problems. Based on those factors (and taking into consideration future needs and potential pressures), regional policy statements will develop strategies for sustainable resource management and identify priority issues and responses. They are statements of policy only; the regulatory measures required to implement them will be generated separately.

Regional plans are optional, except for coastal plans. They will deal with specific resource issues requiring more detailed policies, and can provide the regulatory power to implement regional policies. The regional authorities can use plans to deal with such issues as regional land use effects, soil conservation, water conservation and quality, and pollution discharges.

District government units are based on communities and their surrounding areas, and each one is required to promulgate its own plan. District plans will have policies relating to the integrated management of the effects of land use, subdivision, the control of noise emissions, and the effects of activities on the surface of water in rivers and lakes.

Streamlining the permitting process was considered a key element of the RMA. If a landowner is not operating, or is not able to operate, within the standards established for water quality, air quality, waste disposal, soil management, and so on, he or she will be required to get a special permit, and will have to go through a public process involving an assessment of all the effects of the activity.

However, the permit process will now in most cases be simpler and quicker than it previously was, because now there is only one standard permit process and a standard time limit to that process. Only the local and regional councils will issue permits, rather than a whole host of small boards. If permits are needed from both councils for one property (if, for example, a factory needs both permission to build at the site and permission to discharge contaminants into a stream), there will only be one combined hearing process for both.

The fact that the permit process has been simplified does not mean that it will necessarily be easier to *get* a permit; the sustainability of the proposed action is always the bottom line.

Monitoring, Enforcement, and Appeals

Monitoring is an important part of the RMA. Some of the laws replaced by the act contained no provisions for monitoring, which led to regulation without any real idea of whether or not the regulation was needed or did what it was supposed to do. The RMA requires the gathering of information related to sustainability and monitoring of the state of the environment. It also requires monitoring of the suitability and effectiveness of any policy statement or plan, and of the use of resource permits that have been granted.

Prior to the RMA, the government had already established a Parliamentary Commissioner for the Environment as an independent auditor to monitor the effectiveness of the country's resource laws and institutions. Since its inception in 1987, the commissioner's office has presented numerous reports to the House of Representatives, to parliamentary committees, and to public authorities. It has also published a summary of its findings for the years 1987 to 1991.

This type of monitoring is not related to enforcement. Rather, it involves knowing why objectives have been set, what ends or results are expected (and when), and also checking to see that the methods chosen to achieve those objectives are still relevant and that the costs of achieving them are still worthwhile. The information gathered from monitoring is required to be made public under the RMA.

There are provisions for enforcement of the RMA, of course, and the penalties prescribed for violations, as laid out in the act itself, can be quite severe. Directors of companies that are not in compliance may be liable, and penalties in some cases can even involve jail terms.

The planning tribunals, or environmental courts, are courts of appeals regarding the planning and implementation of the RMA. The environmental courts get involved at two levels. One is the policy-making level: the courts can rule on whether or not a particular plan or policy is in compliance with the requirement to promote sustainable management. The courts have the power to make an independent judgment on what does or does not constitute sustainable management, and can go against the judgment of a municipal or regional council. Because the RMA is in its infancy, the courts have not yet had much experience with this issue, but in time case law will accumulate to define it.

The other level on which the courts operate is that of permit-granting and the enforcement of standards. They can determine whether or not a particular permit was granted in violation of the sustainable management requirement, or denied when it should not have been. They have the power to require permit holders to comply with the terms of their permits, or to meet different standards if circumstances change. Even those who have not been required to obtain a permit can fall under the courts' jurisdiction, if their actions run contrary to the performance standards set by the relevant plans. Any person can request the courts to take these enforcement actions.

Environmental court judges are chosen solely for the purpose of making environmental decisions and are appointed for life, so they have the opportunity to acquire a considerable amount of knowledge about the environment. Consequently, a judge rarely makes environmental decisions on matters he or she knows nothing about. The courts also have members who are not judges, but come from different sectors of society.

Programs Outside the RMA

The RMA is the backbone of New Zealand's overall environmental program, designed to structure a large number of decisions, but it is not the *only* part of it. There are special environmental programs that fall outside of the scope of the RMA. Some were created before the RMA and the government wished to continue them; others required a more detailed handling than could be accommodated by the RMA. They include a program on climate change, one on hazardous substances, and another dealing with waste management. The latter project, which has a strong emphasis on waste minimization and on clean technology in particular, is being done principally through cooperation with industry, beginning with the packaging industry.

The legislation covering mining activities was also dealt with outside the

RMA, in large part due to the fact that the Crown owns most naturally occurring mineral resources. Under the new legislation that has been written in this area, the Crown remains responsible for the granting of mining rights to companies, but the environmental effects of that mining now fall under the jurisdiction of the regional and local authorities, which have the authority to grant permits and set conditions.

These programs all accord with the RMA's purpose of sustainable management, but while some of the measures used to implement them operate within the RMA framework, others do not. For example, the climate change program might make use of the part of the RMA that deals with standards for air quality, but beyond the RMA standards, the program will have its own target of maintaining 1990 levels of carbon dioxide emissions by the year 2000, and a variety of additional measures will be introduced to achieve that.

The Department of Conservation was also created independently of the RMA. Its specific responsibilities are to preserve and protect indigenous species and habitat on government land and to advocate their protection on private land. However, the RMA would also come into play if someone wanted to develop land that was important indigenous habitat; because of the act's provisions, they would probably not be allowed to do so.

Progress toward Sustainability

Although the RMA has not yet been fully implemented, New Zealand, through this and other governmental reforms, has already made some progress toward the idea of sustainable management. The preservation of native forests is one example.

At the time the RMA was passed, the government found that 35 percent of New Zealand's original forest was still standing, and that by focusing on intensive tree farming it could afford to put the remaining old growth in permanent preserve status, which has been done. The government no longer looks at the forest as simply a wood resource, but as an entity that has value of its own. Therefore, the wood that people take from it has to be taken in a way that observes the principles of eco-sustainability, even on private land.

As mentioned above, the farming industry has also made great leaps toward sustainability. One reason is that the country simply stopped subsidizing agriculture. This action dramatically cut environmental degradation, bringing to an abrupt halt many damaging practices, like the application of pesticides and fertilizers. Existing irrigation subsidies were also removed, and as a result, they no longer have any new mega-irrigation schemes.

But removal of the subsidies also completely restructured the farming industry, because of its dramatic effect on people's ideas about what farming was and how to do it. Farmers began to realize that they were on their own and could no longer assume the government would save them if they ran into difficulty. They realized their businesses must be better managed, in a way that fits with nature.

And, as mentioned earlier, they have found sustainable practices can also be quite profitable. They believe that they can do better by specializing rather than by going for a mass market, which has long been the tradition in New Zealand. Now, instead of providing large quantities of lamb and wool to export to England, they are emphasizing quality and targeting discerning consumers who want "clean, green" products.

Looking Ahead: The Transition Period and Future Requirements

A transition period of five years was built into the RMA, to give the government time to develop and implement the necessary standards, policies, and practices. Most of the regional policy statements had been released for public comment by the time of this writing, and the minister of conservation was reviewing the national coastal policy statement. Other standards and policies will follow. All the regional policy statements, and some of the district plans, should be completed by 1996. Even after the transition period has passed, however, the RMA will need to be tested and fine-tuned.

Regional policy statements are intended to have a ten-year life, but their horizon is expected to go beyond that, to at least twenty years. Lindsay Gow, deputy secretary of New Zealand's Ministry of Environment, sees policies and plans under the RMA as "rolling sustainable management programs, updating issues continuously while having an ever-extending outer horizon."

New Zealand's program is behind that of the Netherlands in terms of its information base. However, it has established a program to improve the environmental information used in the decision-making process, and will also use this information as a tool to set up reporting on environmental quality. As the government starts to get more information about the state of New Zealand's environmental quality, it will know better what the targets should be.

According to the above-mentioned report from the Parliamentary

Commissioner for the Environment, in the future the government will need to take action on a number of specific items in order to move toward sustainable management in all resource areas. Gaps noted by the commissioner include the lack of government policies on certain issues involving energy conservation and energy efficiency, and sewage treatment and disposal. The government also needs to develop an overall strategy to improve public transit.

Another potential weakness of the RMA is that it does not in itself deal with international issues and policies, in contrast to the Netherlands' NEPP. One of the reasons for this is context: The Netherlands is surrounded by other nations, and so has much more interaction with them on a regular basis. Their economies are linked, and their pollution shared. The Netherlands therefore has a much more outward perspective than New Zealand, which is an island 600 miles removed from any other country, and so tends to view itself in isolation. New Zealand will come to have a much more explicitly outward view, but even now it has policies designed to provide assistance to developing nations.

In October of 1994, the government announced the "New Zealand 2010" program, the next step toward sustainability by the year 2010.[7] New Zealand 2010 will fill many of the gaps mentioned above. It is specifically an environmental strategy, whereas most of the government's strategies to date have been principally economic. While the country's focus on the principles of the market economy is designed to meet economic needs, 2010 is designed to establish and maintain environmental quality for future generations.

2010 sets a vision for the future, articulating the values and philosophical principles New Zealanders want to have embodied in their environmental strategies. It also involves the setting of the ecological bottom line, which will establish strict limits to pollution and to the use of natural resources, based on the best scientific and technical information. For example, all forests are to be managed strictly on a sustainable basis – no harvest can exceed the growth rates of the forest. Nonrenewables will be managed carefully, and a major policy thrust will be to seek alternatives to nonrenewables.

As part of the 2010 program, the government will:

- weave environmental policy into economic and social policy.
- establish a coherent framework of laws. This will particularly link 2010 to the RMA.
- sharpen the policy tools needed to carry out the program. For instance, market mechanisms will be used, but not emphasized.

- build up the information base regarding environmental quality. This will help establish indicators and determine environmental costs, which will be particularly useful to decision makers and researchers. With this base, a full-cost pricing policy can be implemented, which will ensure that subsidies do not creep into and manipulate the management of key natural resources.
- involve people in the decision-making process.

New Zealand 2010 also sets nine goals for managing environmental quality, similar to the goals set by the Dutch in their NEPP. This type of environmental management by long-term objective has become a key aspect of green plans. It allows them to manage the entire spectrum of environmental and resource issues, with all their interconnectedness, and yet not become overwhelmed by the complexity. New Zealand's nine goals are:

- protect biodiversity
- control pests, weeds, and diseases
- control pollution and hazardous wastes
- manage land resources
- manage water resources
- establish sustainable fisheries
- manage the environmental impacts of the energy sector
- work toward stopping the causes of climate change
- restore the ozone layer

Finally, 2010 includes a review process that will assess the programs' progress toward these goals. The planning and review process will involve a five-year cycle and a one-year cycle. Every four years there will be a progress report to the people of New Zealand.

•   •   •

Unlike such industrialized nations as the Netherlands, New Zealand has the luxury of focusing more on the management of resources than on cleaning up pollution. As a result, it has been able to come up with the better philosophical framework of resource management.

New Zealand's ideas may be more helpful to developing nations than the Netherlands' example. Many of the developing nations tend to see development only as technical and industrial growth, but the New Zealand example of harnessing systems to produce economic returns while maintaining strong environmental quality standards may be far more effective from a number of standpoints, particularly when a nation does not have the resources for heavy manufacturing.

The major difference between New Zealand's program and those of the Netherlands and Canada is that the framework of the RMA was designed not just to affect the way government operates, but also the actions of every citizen, so that all the decisions people make respecting the land, water, air, or coast fit within its framework. Sustainable management fundamentally influences every resource decision far more than any other existing green plan.

The RMA is also distinguished by its coordination of objectives. Sustainable management of resources is the primary drive within every framework, whether it is economic growth or something else. One is part of the other, but, as Lindsay Gow says, "The environment is the top line and the bottom line."

6

Canada's Green Plan:
Making a Virtue of Necessity

If any nation in the world could ignore its environmental problems and survive, it would probably be Canada. With vast forests and fertile prairies, a multitude of lakes and rivers, ocean fisheries, and oil and mineral wealth, Canada is a country rich in resources. Its small population, spread over such a large territory, puts less of a burden on those resources than that of most industrialized nations.

Canadians have an almost mystical attachment to their natural environment; love for the land is part of their national identity. Canada's green plan was born of the citizens' profound desire to improve and safeguard environmental quality in their nation. Courageously, they have moved ahead of most of the rest of the world in their attempt to achieve this goal.

One of the highlights of watching the various green plans develop has been the pioneering attitude of the leaders responsible for them. Canada's team was particularly refreshing in this respect: When it came time to get started on the program, they plunged right in, understanding that they were going to have to do it alone, at least initially. They knew that the plan did not have to be perfect, that it was most important to get a start on it; changes could be made later.

It may have been that the Canadians had a certain innocence because they really did not have to act. Instead of being forced to move ahead because of the pressures from their environmental problems, they were able to take it on as a joyful foray, based on principle.

The Canadian program was actually the result of a process that had been building up for a period of several years. Concern for the environment has been steadily increasing in the minds of Canadians in recent years, as it has in the citizens of other countries. In the years leading up to the green plan, they expressed increasing dissatisfaction with the way their government was handling environmental issues. At the same time, the international movement toward environmental recovery was gathering tremendous momentum. Canada had positioned itself as one of the world leaders on the subject, and it was natural that it should be one of the first to move toward reform at home.

The Canadian public felt that the publication of the Brundtland Report, which marked the emergence of sustainable development as a concept and of a north-south dimension to international environmental issues, required that their country straighten out its own affairs as an example if it were to be effective in helping others, particularly developing nations.

It was in this atmosphere of strong public support for a change in environmental policy that the green plan, officially known as Canada's Green Plan for a Healthy Environment, was born. Philosophically, the green plan is one of the strongest statements of national purpose ever developed, committing the entire country to a process that will fundamentally change its interaction with the environment. As a government and a people, Canadians believe they have no other choice. As Robert Slater, an assistant deputy minister with Environment Canada, has said, it is up to all of us to make a virtue of that necessity.

The Canadian Environment

Canadians' deep historical and cultural ties to the land have instilled in the nation a strong sense of stewardship. Its citizens are well aware of the unique qualities of their country's natural environment, from its abundant wildlife to its rich biological diversity, and know they are dependent on it for much of their wealth and quality of life. Unfortunately, this awareness has failed to stop nonsustainable, exploitive logging practices from threatening portions of that environmental heritage. The recent negative publicity from this practice has been disastrous for Canada's leadership on environmental issues in the eyes of many around the world.

Like other western nations, Canada has become increasingly industrialized in the last century. Although its resource base is relatively pristine in comparison to other industrialized nations, development has taken its toll. For instance, hydroelectric projects have had considerable impact on the environment there, with serious social implications for Canada's native people.

Disappearing species and habitats are a problem in many parts of Canada.

Important ecosystems such as wetlands, native grasslands, and old-growth forests are particularly at risk; on the prairies, more than 90 percent of the original grasslands have disappeared.[1] Unique areas like the Fraser River Basin (home to the world's largest salmon run) and the Great Lakes (the largest freshwater system in the world) have been seriously damaged by pollution. A number of the country's Atlantic harbors have also been seriously degraded by sewage, industrial waste disposal, and shipping spills.

One of Canada's most important ecological responsibilities is the preservation of the Arctic. While most people think of the Arctic environment as untouched, there is in fact evidence of toxics there, including polychlorinated biphenyls (PCBs), dioxins, pesticides, and heavy metals, many of which come from southern Canada and other parts of the world.

Other concerns and potential problems for the Arctic include hazardous waste sites, hydroelectric development, and the industrialization of northern watersheds, which could bring pollution from such sources as pulp mills and oil sands. The exploitation of oil and mineral resources could also lead to significant environmental degradation.

Canadians' awareness of global environmental problems has been increased by airborne contamination in the Arctic and by the numerous transboundary pollution problems they share with the United States. More than half of the acid rain deposition in eastern Canada originates in the United States, as well as a substantial proportion of the smog in some Canadian cities. The United States has also been the major contributor to Great Lakes pollution. Likewise, Canada contributes very little to the problems of global warming and ozone layer depletion, but will suffer from the consequences of both.

The Green Plan Process

Two factors prompted the government of Canada to develop its green plan: principle and politics.[2] The principles, as laid out in the Brundtland Report, were simply that the world cannot continue on its current path. The political motivation came during the election campaign of 1988, when Prime Minister Brian Mulroney saw opinion polls showing how important the issue of environmental degradation was to the public. He gave a number of speeches on environmental topics, and in those speeches he committed his administration, if reelected, to giving environmental factors equal weight with economic factors in decision making. He was reelected, and the government set to work to fulfill those promises.

The green plan creators started from the premise that the plan should be comprehensive, rather than use the issue-by-issue approach the government had taken in the past and which had not worked particularly well. For example, when they had previously tackled the problem of pollution in the Great Lakes, they had come to realize that it was not just a matter of municipal and industrial effluents being discharged into them. They found that much of the pollution in the Great Lakes comes from the air, often traveling thousands of miles before settling in the lakes. In addition, a significant portion of the pollutants comes from runoff associated with land use – agricultural practices and the use of nitrogen fertilizers, for instance. So the lesson was that pollution of the Great Lakes is not a problem of water pollution, air pollution, or land pollution alone, but rather an ecosystem problem, which must be understood and managed accordingly.

The Canadians realized that this lack of a comprehensive approach had caused them to go about creating policy in the most inefficient and costly way. The alternative to running from crisis to crisis was to develop a larger approach that would clean up current problems and also put into place a course of action that would move the country toward sustainable development.

Once the decision was made to go ahead with a comprehensive approach, the environmental ministry, Environment Canada, began working at a furious pace to draft a plan within the time frame the government had set. In March of 1990, the ministry issued a document that would serve as a basis for public consultations; they held their first public meeting on 14 April, and over the next ten weeks conducted some sixty public meetings.

The meetings were held across the country in all the major urban areas of Canada, and in many of the more remote parts of the country, because the government wanted to involve the native peoples, who live in the more remote and isolated areas. These were "multi-stakeholder" discussions, with all the interested parties sitting around the same table discussing the issues. They involved industry, provincial governments, churches, environmental groups, labor, native peoples, youth – everyone who had a stake in what was happening.

Unfortunately, the whole process was a bit too hurried, and that has become something of a problem for the government, because the public did not have enough time to assimilate the information and respond to it. Representatives of some environmental organizations, including a Canadian Sierra Club

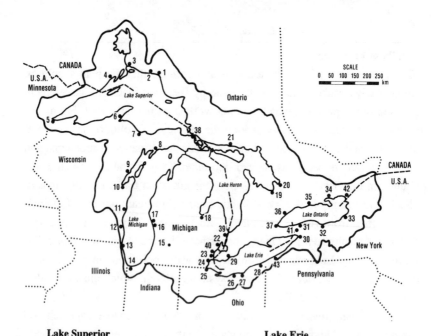

Lake Superior

1   Peninsula Harbour
2   Jackfish Bay
3   Nipigon Bay
4   Thunder Bay
5   St. Louis River/Bay
6   Torch Lake
7   Deer Lake/Carp Creek/Carp River

Lake Michigan

8   Manistique River
9   Menominee River
10 Fox River/Southern Green Bay
11 Sheboygan Harbor
12 Milwaukee Estuary
13 Waukegan Harbor
14 Grand Calumet/Indiana Harbor
15 Kalamazoo River
16 Muskegon Lake
17 White Lake

Lake Huron

18 Saginaw River/Bay
19 Collingwood Harbour
20 Penetang Bay to Sturgeon Bay
21 Spanish River

Lake Erie

22 Clinton River
23 Rouge River
24 River Raisin
25 Maumee River
26 Black River
27 Cuyahoga River
28 Ashtabula River
29 Wheatley Harbour
30 Buffalo River
43 Presque Isle Bay (Erie Harbour)

Lake Ontario

31 Eighteen Mile Creek
32 Rochester Embayment
33 Oswego River
34 Bay of Quinte
35 Port Hope
36 Toronto Waterfront
37 Hamilton Harbour

Connecting channels

38 St. Marys River
39 St. Clair River
40 Detroit River
41 Niagara River
42 St. Lawrence River

Map 2. Areas of Concern in the Great Lakes Basin. As this map indicates, the United States is the major contributor to Great Lakes pollution, but Canada must deal with the effects as well. Source: Environment Canada's Canadian Wildlife Service and Colborn *et al.* (1990). Reproduced by permission of the Ministry of Supply and Services Canada, Government of Canada, 1993.

staffer, attacked the green plan on the day it was released, and have continued to be critical, which has had a negative effect on its public reception. By contrast, New Zealand's process went on over a period of a couple of years, and that country's Resource Management Act has greater public understanding and support today because of it. The Canadian government is now trying to involve the public more extensively, but might have saved itself some problems if it had done so from the beginning.

Nevertheless, the public sent a number of clear messages to the government during these consultations. One of the clearest was that environmental problems needed to be approached in a comprehensive manner – a conclusion that the government had also reached in its own deliberations. People wanted action on a broad range of issues, from global warming to toxics to waste management, rather than a focus on one issue or set of issues. They also placed a high priority on actions that would solve a number of environmental problems, such as the need for more and better information and education.

By the end of this public hearing process the government team had received hundreds of recommendations. It synthesized them into one document, which was submitted to the public for final consultations during a three-day session. More than five hundred recommendations were made by the public during this session; four hundred were incorporated into the plan.[3]

The plan was then submitted for consideration by the cabinet, which took several months to review it. After hundreds of hours of cabinet discussion and a great deal of vigorous debate by ministers, the document was released in December of 1990. The final green plan provides the comprehensive framework upon which Canada will build its environmental policy in the future. It covers a period of five years, and details more than one hundred specific initiatives to be implemented over that period. The government has shown the seriousness of its commitment by pledging $2.2 billion in funds for a six-year period.[4]

The Canadian green plan gains strength from the fact that it was created and adopted by a conservative government, because it has support across the political spectrum. It probably will not be abandoned by future governments, regardless of their political orientation. However, a change in political administrations is always a source of concern for an undertaking as extensive as a green plan; a new government may choose not to make it a priority, allowing it to flounder. It will be interesting to observe what happens in Canada now that a liberal government has replaced the conservative administration that created the green plan, and to what degree and in what form the green plan will be sustained.

As of the date of this publication, the new Canadian administration had yet to be heard from, although the minister in charge of the green plan has reportedly said that it has suffered from a lack of tender loving care. Instead of attacking it, the administration appears to be saying that it will study the green plan and possibly improve it. However, we still do not know how it will rank in the priorities of the new government.

Basic Principles

While the Brundtland Report's vision of sustainable development was a source of inspiration for Canada's green plan, the government realized that the concept itself was too ambiguous to be useful. Accordingly, it developed a sort of statement of purpose for the green plan that articulates what Canadians want to accomplish and what they must do in order to accomplish it:

> While Canadians accept the merits of sustainable development, we understand it is a philosophy, not an action plan. Canadians themselves must determine their own actions for harmonizing our environment and our economy. Sustainable development is *what* we want to achieve. The Green Plan sets out *how* we are going to achieve it together in the years to come.
>
> The Green Plan is not the solution to all our environmental problems. There is no simple solution to the problems we face. No single person, group or level of government has all the answers.
>
> The Green Plan recognizes that, while governments have responsibility to provide leadership, only society as a whole can produce the changes we need to meet the economic and environmental challenges of the 1990s and beyond. This is a national challenge requiring the individual and collective efforts of all Canadians. It will require changes in our thinking and our actions.
>
> It is a plan based on assumptions about the world, the economy and the priorities of Canadians. Like any plan, it was developed knowing that conditions and priorities will change, and new information will alter our assumptions. The Green Plan is designed to change too.[5]

Beyond this statement of purpose, the plan also articulates a number of fundamental "principles for environmental action" as the basis for all its efforts. First is respect for nature, which simply means understanding the intrinsic value of nature and accepting responsibility as its steward, protecting it for current and future generations. It also means adopting the precautionary principle, which means that if we are unsure of the damage certain activities might do, we should err on the side of caution and protect the environment.

Next is the principle of an economy-environment relationship, which states that the well-being of Canada depends on the health of both environment and economy. Included under this are such principles as creating and applying fair and efficient regulations, allowing industry flexibility in meeting environmental goals and targets, viewing the environmental challenge as an economic opportunity rather than a constraint, and investing in science, education, and technology.

The principle of efficient use of resources states that humans must value resources at their true worth and learn to use them frugally. This includes not exploiting renewable resources more rapidly than they can replenish themselves, counting the true costs of depleting nonrenewable resources, and requiring that those who cause the pollution pay to clean it up. Under the principle of efficient use, emphasis is given to reducing consumption, recycling and reuse of products, and recovering of resources.

The principle of shared responsibility simply states that environmental quality recovery is a duty that must be shared by all segments of society and all levels of government, both national and international. Cooperation is an important component of this principle. The leadership principle is the Canadian government's pledge to its people that it accepts the leadership role they have demanded from it in regard to environmental issues.

In a principle called informed decision making, the Canadian plan articulates the belief that wise resource decisions cannot be made unless we know and understand the physical world, its ecosystems, and the relationships between the environment and the economy. This principle stresses the importance of high-quality environmental science, education, and information, and also of a wide cross section of public input into decision making.

The final principle emphasizes the importance of adopting an integrated, ecosystem-based approach in dealing with environmental problems – one that takes into account the vast complexity of environmental interrelationships.

95

Goals and Actions

In order to establish what Canada's environmental policies should be, the green plan's authors, in consultation with the public, developed this list of "priority objectives for Canadians" that would lead the country toward sustainability:

- clean air, water, and land
- sustainable use of renewable resources
- protection of special spaces and species
- preserving the integrity of the North
- global environmental security
- environmentally responsible decision making at all levels of society
- minimizing the impacts of environmental emergencies[6]

They then created action plans for the priority issue areas, with targets and schedules for meeting those targets. More than one hundred important initiatives will be implemented in these areas over the next five years. Where they were not able to identify specific targets, they tried to come up with a process that would eventually allow them to do so.

There are two main thrusts behind these policy objectives; one is aimed at achieving specific resource goals, such as reductions in emissions, while the other focuses on changing behavior.[7] Both the Canadians and the Dutch have sailed off into uncharted waters on the issue of changing behavior.

The government realized that if Canada were to achieve sustainability, the public would have to understand how environmental problems are caused by behavior at all levels of society. Implementing sustainability requires changing the way decisions are made, from the institutional level on down to the individual, in order to reflect the true value of environmental resources. Environmental issues and considerations need to be thoroughly integrated into the processes used to reach decisions, not ignored or dealt with as an afterthought. This is true whether it is a government deciding on an economic policy or an individual choosing what sort of transport he or she will use to get to work.

If this process of change is to work, everybody has to be involved – industry, government, institutions, and most importantly, every individual in all of his or her roles, as commuter, tourist, consumer, and so on. But in order for this to happen, government has to lead the way, not just by setting an example, but through a whole series of actions designed to get people involved.

One strategy is to present environmental recovery as a challenge, calling upon people to work together and to believe that their actions will make a difference.

In fact, Canada's green plan does issue a challenge to the nation in the form of a series of assertions: "Canadians *can* make better individual and collective decisions"; "Canadians *can* clean up the mistakes of the past and ensure they do not reoccur in the future"; and so on.[8] However, they did not make use of an idea that both the Dutch and the New Zealanders have found useful in terms of motivating their citizens: concern for future generations. The planners did not put much emphasis on it because the country's leadership did not, but if they had they would probably find, as did the Netherlands and New Zealand, that this is a concern the public responds to very strongly.

However, even people who are motivated to change cannot do so if they do not know what will help. The Canadian government has set out to develop partnerships with important sector groups to work out the best and most efficient ways to achieve their goals. They intend to work with industry, environmental groups, native people, youth, communities, educational and research institutions, and other groups. They also have to develop consultation practices and procedures that will allow them to do this.

Toward this end, they have created national and provincial Round Tables on Environment and the Economy. The round tables attempt to bring together such traditionally adversarial groups as industry, environmentalists, and government. Because they operate outside the established structures of government, they have had some success in making recommendations that cross traditional lines and break out of political gridlock.[9] However, because round tables attempt consensus in their decision making, they have encountered some problems. One is that achieving consensus takes a great deal of time – so much time, in fact, that policy makers cannot effectively participate; they have to send assistants in their place.

Provincial round tables such as those in British Columbia, Ontario, and New Brunswick have developed sustainable development strategies with recommendations for their provincial and local governments. However, the round tables in some regions are declining – a number, including British Columbia's, are no longer funded and are being allowed to die out.

The National Round Table does excellent studies, and is working in thirteen program areas, from forestry and fisheries to trade education policies. It also publishes books, sponsors conferences, advises the government, and brings together major stakeholders to negotiate agreements. It also oversees, in conjunction with other national groups involved in sustainable development, Canada's follow-up to the Rio Summit, a national stakeholder process called the "Projet de Societé."

Round tables are currently being explored as an option in other countries, including the United States. It is important to remember that, while they can help increase public involvement in environmental policy making, round tables are not themselves green plans. Round tables are public discussions regarding issues; a green plan is a funded policy plan for action, with a majority of the parties committed to the goals and to finding ways to achieve them.

The government is also building partnerships with citizens and their organizations by funding projects that involve communities in environmental activities in their area, creating an eco-label program for consumers, and assisting indigenous communities in developing their own environmental action plans. And it is also leading the way in terms of changing its own behavior: the green plan is "a comprehensive plan of action for the entire Government of Canada, supported by more than 40 federal departments and agencies,"[10] a government-wide initiative showing government-wide commitment.

The proposed Code of Environmental Stewardship and Canadian Environmental Assessment Act would legally obligate the government to integrate environmental considerations into all its actions. Canada already conducts environmental assessments of all proposed program and policy initiatives; it is one of the few nations in the world to do so.

The federal government has made some institutional innovations that have given environmental concerns a far greater role and share of power in government. The Minister of the Environment is now a member of the most powerful cabinet committees that exist in the federal government. There is a cabinet committee on the environment, and also a Canadian Council of Ministers of the Environment, which is a forum that brings the federal and provincial ministers of the environment together on equal footing.

The federal, provincial, and territorial governments are also attempting to forge a new relationship based on cooperation and sharing of information. The objective is to ensure nationwide consistency in terms of goals, standards, legislation, policies, and programs.

Neither the public nor government officials can make sound decisions unless they are educated and well-informed. Accordingly, the government has made public consultation, access to information, monitoring, and assessments priorities of the green plan. It requires annual reviews of the green plan in consultation with the public, as well as regular state of the environment reports. Researchers are working to develop environmental indicators.

Many green plan initiatives are aimed at improving the information flow to decision makers in all of their institutions. The government is also working to develop specific monitoring systems and environmental audits, in order to assess the effectiveness of particular policies and programs.

The green plan's other policy arm, targeted toward specific resource issue areas, is more like a traditional environmental plan. It covers all the issues in which resource agencies are usually involved: air, water and land pollution issues, issues of sustainability in forestry and agriculture, national parks issues, environmental emergencies, and so on.

The actions to be taken are far too many to list here, but they range from committing the federal government to reducing its own waste by 50 percent by the year 2000, to requiring new regulations on dioxin and furan emissions from pulp and paper mills.[11]

Since the announcement of the green plan, the government has started work on thirty-six major initiatives designed to create a healthier environment.[12] Some of these initiatives are national in scope, such as the Federal Waste Reduction plan, the Pulp and Paper Regulatory Package, the Canada/U.S. Air Quality Accord, and the Research and Training in Environmental Studies Program. Other initiatives focus on specific regions, such as the Arctic Environmental Strategy, the Fraser River Basin Action Plan, the Great Lakes/St. Lawrence Pollution Prevention Plan, and Remedial Action Plans for twelve Atlantic harbors. In order to achieve the objectives laid out in the action plans, the government will use a combination of legislative, regulatory, and market tools. Some of these regions, such as the Fraser River Basin, are already moving ahead on strong management plans.

Three years after its launching, the Canadian green plan is moving along, but not as dramatically as was hoped. The economy has continued to be a difficult issue. Although funding remains substantial, it must be stretched further than was originally planned.

One gets the sense that while government agencies in Canada understand clearly where they are going to go with the green plan, the person in the street does not – and is not linked into it as a national purpose. That is most likely a result of the hurried consultation process, the over-emphasis on funding rather than content, and the negative rating of the prime minister who launched the program.

In the Netherlands, the first message to the public was one of inspiration and challenge from the Queen, who communicated the importance of behavioral change. A similar situation occurred in New Zealand, when an incoming prime minister responded to the former prime minister's emphasis on economic development by saying that there is more to life than that.

The message here is that it is better to emphasize principle than money – or to put enough of the budget into education to maintain awareness on the part of the public. In Canada's case, the focus has shifted to what the money is being spent for; people mistakenly see it as a pork-barrel issue, instead of understanding the importance of the principles and the approach. The passion of the people is the same, however: in the polls, most Canadians remain deeply concerned about the environment.

There are other things besides the financial emphasis that the Canadians may want to revise. For example, although the plan pulled together all the money and all the different departments of the federal government, it did not really spell out the provinces' share of commitment and responsibility. This could prove important, since many environmental issues fall under the jurisdiction of the provinces.

Another important factor in the Canadian example will be the future political leadership. At the time of this writing, the Canadians had just gone through a change in national leadership, and that will always be a critical juncture for long-term plans like these. Nonetheless, the politicians are aware that the public rates environmental quality very highly, and whoever is in office is likely to continue the effort.

Another factor slowing the green plan has been the outspoken criticism of some in the environmental community. Some of their criticisms are not without cause: the exploitive clearcutting practices being allowed in its forests have hurt Canada's image in the world's eyes over the last several years, and have even affected people's attitudes about the potential of the green plan. However, it is important to keep in mind that the green plan is a process, and that forestry practices are just one part of the whole package. Tremendous good has come from the green plan, and there is potential for much more in the future. In fact, British Columbia, where some of the worst logging practices seem to go on with unrestricted abandon, has made some major improvements in the past year. Nonetheless, it is sad that much of the good work that is being accomplished in Canada has been undermined due to exploitive forestry practices.

Canadians have provided a great deal of world leadership in environmental affairs, a fact that has tended to get lost in the storm of criticism. Nonprofit leaders, as well as government officials such as Jim MacNeill and Maurice Strong, were very involved in the process that led to the UN's environmental advances in Stockholm and Rio. MacNeill, a member of the Brundtland Commission, is himself a severe but measured critic of the Canadian green plan; he says that while Canada is better off for having the plan, it will not by itself halt the country's environmental decline. Nonetheless, he believes that the people will demand that the government stay on the green plan course and deliver environmental improvement.

The Canadian green plan was a rare case of a society deciding to radically change itself in order to meet a common goal. It is unique in its strong philosophical base; in the open, democratic process within which it was created; and in its extraordinary optimism. Time is now the primary issue: progress is slower than observers might have liked. But the green plan is alive, and its programs and policies continue.

The spirit of the plan can be summed up by two quotes. The first is from the plan itself:

The challenge *can* be met. But it will require a fundamental change in the way we use the environment in our pursuit of economic growth. Change takes time: centuries-old values and attitudes are not transformed overnight. The task will be made easier as people become aware that, like any natural species, our success depends on our ability to adapt to our environment. The environment will not adapt to us.[13]

The second is from Robert Slater of Environment Canada:

I believe that green plans are unavoidable and inevitable.... There are two ways that you can look at this inevitability. You can look upon it as an imposition from outside; you can look upon it as making you a victim, putting a burden on you. Or you can look upon it as making you a winner, creating for you a terrific opportunity....

Sustainable development, I believe, represents a huge opportunity for those people who approach it as winners, as a great strategic opportunity to see themselves and their country into the next century. I believe that it provides a basis for unity within countries and amongst peoples that has previously eluded us.[14]

7

On the Green Plan Path

The policies in a number of countries approach the green plan level, but are not quite there yet. These plans are admirable, far-reaching, and usually better than those of other developed nations, including the United States, but they are not part of the functioning policy and everyday decision making of government. Because they are not integrated, they do not reach from the individual, through the community, the factories and institutions, and the various levels of government, to ensure that the programs achieve the desired result.

The policies of these countries are important, but until they are integrated, they will remain at an earlier stage of management vision than the plans of countries like the Netherlands, New Zealand, and Canada. For the purposes of this book it is useful to review them, but they do not require the same amount of detail.

Because of the actions they have already taken, these "near-plan" countries could at any time pull everything together into a comprehensive plan. For instance, Austria has green plan legislation already written and in front of Parliament, awaiting passage. As in any political process, it may go through instantly or it may take years.

Even a plan that seems to be progressing well could easily falter at any moment, although it would only be a matter of time before it got back on track. The point is that these nations' programs, like those of the green plan countries, can change at any time. The snapshot of them that this book provides is only that: a quick picture of where they were when it was written, in early 1994.

Among the countries that are on the right track, the Scandinavian ones most closely approach true green planning in their policies, yet they have never taken that final comprehensive step. My intent is not to be critical, but to draw the line at a point that clearly denotes what is and is not a green plan. In general, the Scandinavian countries have been out in front of the rest of the world in terms of environmental protection. They have reacted to problems more quickly and effectively than other countries, and so never developed the rather awesome pollution problems that tend to prevail elsewhere. When they discovered problems of air pollution or water pollution, they tackled them immediately, rather than ignoring them until they were too bad to ignore any longer, as has been the custom in the United States.

Of course, Norway and other Scandinavian countries have been drastically affected by transboundary air quality problems, such as a high level of acid deposits that originated in other parts of Europe. Consequently, they believe strongly in the need for sustainability, and continue to implement progressive policies toward that goal. But because they have been more careful in dealing with environmental issues as they emerge, the need for an overall, coordinated plan has not been as urgent for them.

Norway

Norway's prime minister, Gro Harlem Brundtland, headed the UN commission that issued the report *Our Common Future*. Her interest and involvement in global environment and development issues, and particularly in the concept of sustainable development, has had a profound impact on her country's environmental thinking.

Our Common Future, published in 1987, provided the basis for Norway's current environmental policy, which was presented as a report to Parliament in 1988–89. The report attempted to deal with all aspects of environmental protection in the context of environment and development, and was designed to serve as a framework for future environmental policies.[1]

When the Norwegians set out to develop their environmental policy report, they first studied a broad spectrum of resource issues and tried to diagnose the problems of each. They looked at the various sectors of society and their environmental problems, and realized that in order to solve them they needed to create an infrastructure that would allow them to do so. In other words, they would have to increase the number of choices available to people for things like recycling and mass transit.

The Norwegians included in their plan two innovations on the national level of policy. First, they looked at the budget of each of their ministries to see how much each was spending on environment, and what its aims were; this information goes into each year's overall budget statement. Second, they instituted an annual state-of-the-environment speech to Parliament, to be given by the minister of the environment.

On the local level, they established an environmental program for local authorities, to help them come up with their own plans. To this end they have provided a free consultant for one year for about 100 of their local governments, and have raised money to provide help for others to do the same. These local plans will be subject to local priorities and interests, and not just a reflection of the national plan.

The Norwegians set up some very general aims for their plan, such as the precautionary principle – anticipating, preventing, and attacking sources of environmental degradation as well as not using scientific uncertainty as an excuse for inaction – and the belief that critical loads on nature should not be exceeded.

About ten years ago, the Norwegians started shifting their environmental policy toward performance-based regulation rather than prescriptive regulation, with the idea that it is what you do that matters, not how you do it. For example, the government sets emissions limits and performance standards for business, then lets the businesses determine how to meet them. This fits in with Norway's tradition of a small central government that devolves responsibility to local authorities. They believe that it is not the government's job to give detailed environmental advice, but rather to identify the challenges that lie ahead and ensure that there are mechanisms in place that allow people to make the right choices.

Once a year, Norwegian businesses must publish an environmental report that details their resource use and environmental impact. The government works with industry to help them with quality control in their processes, doing things like reducing pollution internally through more efficient processes.

The plan has a chapter on each "sectoral group," similar to the Netherlands' target groups. It sets many specific goals and targets, including limiting carbon dioxide emissions in the year 2000 to 1989 levels, and then trying to reduce them further; reducing sulfur dioxide emissions by 50 percent in relation to 1980 levels (this has already been accomplished); and reducing nitrogen oxides emissions by 30 percent by 1998, in relation to 1986 levels.[2]

Recently the Norwegians have shifted more toward the use of economic measures to achieve their goals, such as raising taxes on environmentally unsound activities and lowering taxes on environmentally beneficial ones. They are introducing toll road strategies; part of these funds will go toward mass transit. They have a carbon tax on fuels, and are applying pollution control laws to roadbuilding projects.

The government has set up a Council for Sustainable Development, which includes business, labor, government, and environmental groups. The Council deals with issues such as whether or not to adopt carbon dioxide taxes and determining the country's international aims. The Norwegians have found the Council useful in terms of institutionalizing sustainable development, and also in addressing the need for collective solutions.

Norway devotes a great deal of effort toward international agreements on environmental measures. For instance, it would like to see the European Union adopt the carbon tax so that its businesses can remain competitive. More stringent Europe-wide controls on emissions would help its acid rain problem.

Norway's accomplishments in the field of environmental recovery are remarkable; one can only wish that the United States were as progressive. The only reason I do not consider their policies a green plan is that they have no formal structure that integrates them throughout government. While the environment ministry has worked well with other ministries on environmental issues, their relationship is not written into law.

It is possible that Norway does not have a strong, centralized environmental policy because its government has a longstanding tradition of decentralization, placing more authority in the hands of the counties and municipalities. There may be another, more political reason, however: any country with a large oil industry – like Norway – may have a very difficult time putting a comprehensive, integrated plan into effect.

In countries where the oil industry plays a large role in the economy, it is usually very powerful in the political arena. Unfortunately, it will often use that influence against a comprehensive resource plan, because it believes that such a plan's provisions regarding energy and fuel consumption will hurt the industry financially. I have been told that that is the case by sources in these countries, and it is certainly true in the United States.

This is probably a temporary obstacle. Most assessments of oil reserves estimate that, based on current rates of consumption, the world's stock of economically recoverable oil will last only another fifty years.[3] A number of progressive oil companies have a fair understanding of the difficulties of energy supply in the future, and are in the process of diversifying. In several instances they have been responsive to such long-term issues. It is unlikely that the Netherlands' plan would have gotten off the ground without at least the quiet support of Royal Dutch Shell, one of the world's largest oil companies, which has moved well out in front of most other corporate structures with its innovation and long-term planning.

Sweden

Environmental protection in Sweden is currently based on five action programs, covering air pollution, marine pollution, chemicals, nature conservation, and freshwater conservation.[4] All were included in an environmental bill presented to Parliament in 1991, which replaced a plan from 1988, which had in turn replaced an earlier plan. The 1991 plan had a goal of integrating and strengthening the country's environmental policy.

In this bill, the Swedes defined four general objectives: protecting human health, preserving biological diversity, managing the exploitation of natural resources to ensure sustainable utilization, and protecting natural and cultural landscapes.

They have also adopted seven principles of environmental protection: respect for human and environmental critical load limits; use of the best available technology, requiring each country to ensure that activities within its borders do not cause environmental damage in other countries; the precautionary principle; the polluter pays principle; the substitution principle (hazardous substances should be replaced by less hazardous ones whenever possible); and the principle that an environmental impact assessment should be done for any activity that may have substantial adverse effects on the environment.

Achievements so far include a reduction in sulfur emissions of about 70 percent; in emissions of persistent organic compounds from the pulp and paper industry of 75 percent; in phosphorus from cities and municipalities of 90 percent; in the use of pesticides for agriculture of 50 percent in the last five years; in the use of cadmium in consumer products of 50 percent. Metal emissions from manufacturing industries are down by 70 to 80 percent; CFCs have been cut by 70 percent in the last few years. Targets set for the near future are equally impressive: the government wants to reduce pesticides by another 50 percent in the early 1990s and the use of chlorinated solvents by 90 percent, among other goals.

Some difficult problems remain, such as leakage of nitrogen from agriculture into coastal waters, and nitrogen oxides emissions. Another problem is that 80 to 95 percent of the acidic airborne substances deposited in Sweden originate in other countries. This is a particular problem in terms of the acidification of their inland waters, soil, and groundwater.

Sweden's cost for environmental protection amounts to approximately 2.5 percent of GDP. Approximately 10 percent of this is financed from budgets of local, regional, and national governments; the rest comes from the costs imposed by regulations, and is paid by households, industry, forestry, agriculture, transportation, and the trade industry, according to the polluter pays principle. A wide range of instruments is used for implementation, including such traditional tools as regulations, ambient standards, licensing of stationary sources, and technical standards.

The Swedes' use of economic instruments as a supplement has increased considerably in the last few years. Their emphasis has been less on subsidies than on financial disincentives. They impose charges on carbon dioxide, nitrogen oxides, and sulfur that are designed to reduce emissions to certain target levels. They also charge excise duties on fossil fuels, pesticides, and fertilizers, as well as on mercury and cadmium in batteries. From 1993 on, new vehicles will have three different environmental classifications: the least environmentally friendly will be more heavily taxed, while the most friendly will get tax breaks.

Sweden has already reduced many emissions far below the standards in other European countries, so the marginal costs of further reductions would be very high. In some cases they might have severe impacts on the competitiveness of Swedish companies and farmers. Because of this, they are looking for joint ventures in other countries that can reduce transboundary air and water pollution in a cost-effective manner. By investing $100,000 in Poland – which is one of the upwind eastern European countries whose pollution is killing their lakes – they can have a tremendous impact compared to what that $100,000 would achieve in Sweden.

In recent years, Sweden has been placing increasing emphasis on the diffuse pollution caused by product use. As a result, it is concentrating its efforts on becoming a more "ecocyclic" society – in other words, adopting what the

Dutch term a "lifecycle" approach to production and consumption. Characteristics of this approach include reducing the use of materials and energy in production, emphasizing product quality over quantity and producing more durable goods, and recycling and reusing as much as possible. One instrument the Swedes use to accomplish this is to make producers more responsible for the recycling and reuse of their products.

There has been a dramatic change in the attitude of Sweden's corporations in the last five years. Where once they were opposed to stringent environmental standards, now many are learning to take advantage of them. This does not include all industries, of course, but a number of big companies, including Volvo, are taking a great deal of interest in environmental development.

Twelve nuclear reactors provide about 50 percent of Sweden's power needs, but are scheduled to be phased out no later than 2010, by public demand. After that the Swedes will be under great pressure to look at energy conservation and alternative sources.

Their eventual goal is to reduce emissions to levels that even the most vulnerable areas of their country can sustain. They are nowhere near that goal, primarily because of transboundary air pollution.

Although Sweden is on the track to a comprehensive plan, it has not quite arrived. With its record of progressive environmental policies, however, it will be only a matter of time before one is adopted.

Denmark

Denmark suffers from many of the same environmental problems as its neighbors in Scandinavia, including acidification, eutrophication, the dispersion of pesticides and other toxic chemicals, threatened coastal waters and groundwater supplies, waste disposal, and the loss of wildlife and native habitat.[5] Also like its neighbors, Denmark developed progressive environmental policies early on, and made great strides in combating these problems in the 1970s and 1980s. Its achievements in the area of energy conservation have been particularly impressive; after the oil crisis in the 1970s, it reduced consumption by 25 percent.

The Danish Ministry of the Environment was established in 1971, and the National Agency of Environmental Protection (NAEP) in 1972. The Environmental Protection Act provides the legal basis and framework for much of the country's environment-related activity. It governs all activities that can pose a threat to the soil, water, or air, and empowers the Ministry of the Environment to lay down regulations regarding them. The NAEP is an administrative body that coordinates environmental activities on a national scale and drafts the regulations to be adopted by the ministry.

In response to the Brundtland Report, Denmark published an "Action Plan on Environment and Development" in 1989. This document focused on cross-sectoral policies to achieve sustainable development, and included more than 150 initiatives. Since then it has developed action plans for a number of key issues, among them acidification, waste management, carbon dioxide reduction, and energy, with specific action and quantitative targets spelled out for each. In 1992, a government publication entitled *Environmental Initiatives in the 1990s* spelled out overall environmental objectives for Denmark in the areas of human health and well-being, quality of life, nature protection, and international cooperation.

Despite the Danes' many environmental policy successes, a comprehensive, integrated, long-range national plan has remained elusive, because local governments have tended to bear the largest portion of responsibility for much of Danish environmental policy. This has made overall strategies for sustainability difficult to develop and implement nationwide. After the Rio Earth Summit, however, the government began to develop a "continuous strategic environmental planning system," and a report it released in March of 1994 indicated that this system will essentially be a green plan.[6]

According to the report, strategic environmental planning means establishing goals and strategies; monitoring actual environmental conditions to assess progress; and building in the means to adjust policy to reflect changing conditions. To this end, the environmental ministry proposes to publish regular progress reports that will present an overall picture of the state of the environment and assess the impacts social behaviors have on it. The report will look at trends, create environmental impact assessments, and develop scenarios and modeling.

The second major tool the ministry will develop is a regular environmental policy statement, or white paper, which will present the government's proposals for environmental priorities, goals, and specific initiatives and actions. This white paper will be comprehensive, covering all environmental issues, and will be based on the information provided by the progress reports. The environmental ministry will draft the targets in consultation with all the key ministries, which will then be responsible for integrating them into their own policies and decisions.

Other green plan elements to be built into Denmark's strategic planning system include the development of a strong database, computer modeling, and environmental indicators; efforts to integrate environmental data into national economic accounting; analyses of industrial life cycles and of consumption patterns; and research into environmental technologies.

After an early start in the field of environmental policy and many successful programs, the Danes for a while fell behind, unable to break away from the piecemeal, issue-by-issue approach. That meant that, even though they had strong policies in many areas, they were unable to make headway on certain types of problems that require the comprehensive approach. Once their new planning system is implemented, they should be able to progress once again.

Austria

During the 1970s and 1980s, Austria accomplished a great deal in terms of environmental cleanup and protection.[7] For example, over the past twenty years Austria cleaned its lakes to the point where they are all of drinking water quality. While there is still much to be done, transboundary pollution is now one of the country's biggest problems. Because Austria is a very small country right in the middle of Europe, it is especially vulnerable to the actions of its neighbors, particularly in eastern Europe.

Austria's early environmental policies, begun in the 1970s, included targeted strategies for emissions reductions, and many of these are now in place. In the 1980s the government decided it needed to take a more long-term, comprehensive approach. As a first step, it is attempting to introduce environmental accounting in industry. It is also in the process of reviewing its accounting system for the GNP, to integrate the costs of resources and environmental damage. In addition, the government monitors progress on environmental issues by requiring that the minister of environment present a state-of-the-environment report to Parliament every two years. Moreover, Austria has developed a waste management plan that is truly comprehensive.

The Austrians are working on other legislation to strengthen their environmental policies; for instance, they are considering imposing a carbon dioxide tax and an energy tax, but will only apply the carbon tax in conjunction with other European Union countries, in order to protect their businesses' competitiveness. Their car prices already include a tax based on fuel consumption, and they are exploring other types of ecological taxation.

Another of Austria's environmental policies is to phase out chlorine in the pulp and paper industry. It is also increasing its forest cover, even though the country is already 46 percent forested – it needs forests in the mountains, or the valleys will be uninhabitable due to erosion and avalanches.

Austrian businesses resisted the changes at first, but now many are leaders in environmentally sound technologies and techniques. The government gave industry incentives to look for locally compatible, small-scale solutions to environmental problems. Although the country's businesses cannot always compete with huge corporations, they are now in a good position to help their eastern neighbors, because their solutions do tend to be small, less expensive, and easily adaptable. Helping its neighbors also helps Austria's environment, of course.

In the area of energy, Austria is turning toward biomass conversion, conservation, and somewhat to solar power. Although it is clean, hydro power generation is a problem because the Austrian public does not want its rivers to be dammed. Austria has no nuclear power plants of its own, but imports energy from France, which does use nuclear power. The government is attempting to move away from this policy, because it is inconsistent with the country's philosophy regarding nuclear power.

The Austrians are now in the midst of developing a green plan, called the Austrian National Environmental Policy Plan (NUP). In 1992, working groups were assigned to develop targets for seven sectors: industry; energy and fossil fuels; traffic and transport; agriculture, forestry, and water management; tourism; management of waste streams, minerals, and natural resources; and consumers. The working groups' interim reports were approved early in 1994, and an initial version of the NUP, integrating the individual sector-specific programs and action plans, was expected by late 1994. The plan will then be submitted to the Council of Ministers for approval.[8]

One of the reasons Austria's green plan was stalled for so long is that there was a great deal of debate over whose facts were right, the government's or the opposition's. This is a problem that comes up again and again; the Dutch managed to avoid it by relying on an independent and highly respected scientific research institute in preparing its plan.

United Kingdom

In 1990, the United Kingdom adopted a white paper called *This Common Inheritance,* a precursor to an actual green plan.[9] The government had for years had a policy that environmental concerns were to be built into policies across the board, but it was only a statement. At that time the idea was new, and the issue was not a major government priority, so it was not widely applied. But prior to the adoption of *This Common Inheritance* the pressure on the government to introduce stronger environmental policies had been growing, both from the public and from other European Union nations. By 1989, the government felt that the time had come for a more comprehensive approach.

Its first step was to set up an audit process: every government agency, including the department of environment, was instructed to review all its policies and issue a report on where environmental progress could be made. The department of environment initiated a consultation process, issuing invitations to environmental nonprofits and other groups to participate. These consultations led to *This Common Inheritance.* The white paper has chapters on European issues, world issues, global warming, land use, countryside policy, urban issues, and heritage issues in addition to the expected issues like air and water pollution, waste, hazardous substances, noise, and so forth. It also contains chapters that cut across traditional environmental boundaries, including one on education, one on institutions, and one each on Scotland, Wales, and Northern Ireland. In addition, the white paper has a statement of principles. These are: stewardship; responsibility toward future generations; fact, not fantasy, as a basis for action (a sound scientific basis and sound economics); more information for the public, to enable them to make well-informed choices; international cooperation; best instruments (markets as well as regulation, etc.); precautionary action where the risks justify it; and the principle that action is for everybody, meaning that the environmental ministry can only do so much on its own.

After publication, the government announced a list of "green ministers" – someone in each ministry who assesses the environmental impact and implications of all his or her department's policies and spending programs. This is the person whom people can contact regarding problems they have with a department's policies. They also kept two ministerial committees that had been set up during the process of creating the plan – an important element in a cabinet system of government. Additionally, they have established permanent forums that include the environmental ministry and local governments, non-governmental organizations, and business.

This Common Inheritance contains a number of specific actions and goals for each ministry to work toward. The ministries must then report on their progress toward each goal, and discuss what actions they are planning to take next.

The white paper itself does not contain much legislation; that is being done through bills presented to Parliament. One of the concepts being legislated is a system of integrated pollution control based on requiring use of the best available technology. The white paper itself did have some targets and time-lines, as did an assessment of the white paper that was issued after the first year. Three annual reports on the white paper have also been issued, documenting progress.

In January of 1994, the prime minister announced the "UK Strategy on Sustainable Development," which is the country's follow-up to the Rio Earth Summit. Among other things, this initiative focuses on the relationship between economics and the environment, looks at expected trends over the next twenty years, and sets up mechanisms for involving the public in the process. It is too early as yet to assess the effectiveness of this initiative.

The primary defect in British environmental policies to date is the lack of a serious commitment of funds – an important measure of a government's commitment to any particular policy. However, many in the United Kingdom continue to think and write and work toward the goal of a comprehensive environmental policy, and their efforts will continue. The ideas and the framework are there, and once the leadership makes environmental sustainability a priority, the country can quickly catch up.

Germany

Like many European countries, Germany made great progress in reducing its pollution emissions in the 1970s and 1980s. Sulfur dioxide emissions fell by 73 percent, particulate emissions by 74.8 percent, and carbon monoxide by 43.8 percent.[10] Energy conservation practices have led to a greatly reduced rate of growth in energy use.

The German government has also focused on water pollution prevention and purification, waste management, and conservation. A great deal remains to be done in these areas, however; the situation in the field of nature conservation and soil protection is described as "critical." Damage to the soil structure and an alarming rate of species loss indicate the severity of the problem.

The biggest environmental challenge Germany faces will be repairing the damage done to the former East Germany. According to the government, the environment there is in some cases in a "catastrophic state." In November of 1990, the German environmental minister presented a clean-up and development strategy for the eastern part of the country.

German environmental policy is to employ a variety of instruments to encourage better environmental behavior, including laws and regulations, market instruments, environmental impact assessments, area planning, voluntary promises by trade and industry, dissemination of information and advice, and environmental education.

The Germans have adopted a number of common environmental principles as a base for all their efforts, including the precautionary principle and the polluter pays principle. Like the Dutch, they focus on production processes and life-cycle management in order to minimize waste. Cooperation is also a key concept for them, especially in terms of the experience and information the states and towns of the west can share with those in the east. The next step, they say, will be integrating ecological requirements into other policy areas, such as transportation, agriculture, and energy.

Germany has some of the most progressive environmental policies in the world in certain issue areas, including one of the best laws on packaging, but its approach is still piecemeal. Like many other European countries, it is looking toward a comprehensive, multi-sectoral approach, but as yet has no framework for one.

Singapore

Singapore, a city-state with essentially no natural resources that has experienced rapid industrial growth in the last few decades, may seem an unlikely candidate for green planning.[11] Yet it may be precisely the limitations of its small, highly developed land area that have prompted Singapore to adopt a comprehensive environmental program. Its highly centralized government, in which all the departments work in concert, has made it relatively easy to do so.

The country's space limitations have forced it to practice careful land-use planning for many years; all land has been zoned under the city's master plan since the 1950s. Environmental impact assessments are required for all new development. Singapore is already known for having some of the world's strictest pollution control standards (its air quality standards are particularly impressive), and a modern, efficient, and cheap mass transit system (as well as strict regulations on car ownership and use). Now it is positioning itself to be one of Asia's leaders in environmental technologies and services.

In general, the government works closely with business, and environmental issues are no exception. When Singapore began to industrialize in the late 1960s, the government decided it wanted to avoid the pollution problems other countries had faced. It focused on preventing them at the source, and applied strict regulations from the beginning. As the country developed, it was able to take advantage of new environmental technologies, both in its industries and in the creation of its infrastructure.

In 1992, Singapore issued the "Singapore Green Plan" as a preliminary step toward a formal, comprehensive plan. Primarily a vision statement, the plan lists broad goals in areas across the resource spectrum, as well as laying out a program of environmental education and information aimed at helping citizens and businesses make better environmental decisions.

Although the plan lists some specific targets, such as setting aside or reclaiming 5 percent of the country's land as open space and setting stricter emissions standards, it is by nature a more general document, outlining broad strategies to make Singapore a "model green city" by the year 2000. A more comprehensive action plan and specific action programs based on the green plan were released late in 1993.

Singapore's green plan does not discuss, for example, the integration of environmental concerns throughout government departments and agencies (most likely because integration is the rule rather than the exception for Singapore's highly centralized government), and the development of new strategies for working with business on pollution prevention. The plan does, however, discuss the idea of the government taking a more proactive role in environmental management by promoting environmental audits and the use of clean technology. Energy conservation, while not covered in the original green plan document, is now an important component of the program.

Possibly the weakest part of the green plan, and of Singaporean environmental planning in general, is its lack of a strong commitment to natural ecosystems and the conservation of native plants and wildlife. Although the green plan does commit the government to setting aside or reclaiming 5 percent of the country for nature, Singapore is urbanizing rapidly, and much of its environment has already been altered.

Singapore's environmentalism could well be a model for countries that are currently in the process of industrializing – and the government clearly intends it to be. The UN has cited it as such, not only for its policies and plans, but for the Environment Ministry's ability to effectively monitor and enforce those policies.

The European Union

The twelve-nation European Union, the political, social, and cultural entity that has developed from the trading bloc of the European Community, has always had a strong economic focus, but it has also made environmental quality a high priority. The latest action program on the environment to be promulgated by the European Commission (the European Union's administrative arm), entitled *Towards Sustainability,* is the best it has yet developed, in many ways similar to a national green plan. This plan, exhaustive and well-written, has a clarity and purpose I have rarely seen in an environmental plan, whether corporate or government. It is an indication that green planning can also work for an association of nations, or an association of states within a nation, such as in the United States.

The European Union has always shown concern for environmental quality, even when, as the European Community, it was primarily a trading bloc. As early as 1972, the heads of member states declared that "Economic expansion is not an end in itself...it should result in an improvement in the quality of life as well as in the standard of living."[12] In the 1980s, the community's main concerns were still focused on the internal market, but in the 1990s it shifted that focus to finding a development path that would be sustainable.

The action plan describes the characteristics of sustainable development as maintaining the overall quality of life, maintaining continued access to natural resources, and avoiding lasting environmental damage. In essence, the document says, this can be translated as "Don't eat the seed corn which is needed to sow next year's crop."[13]

The European Union's decision to take up the challenge of ensuring that growth is environmentally sustainable is a remarkable action for an entity that developed out of a trading bloc. But *Towards Sustainability* stresses the importance of building a business base that incorporates environmental technologies and practices that make it more competitive. The ultimate reasoning, of course, has to do with the economy's dependence on a healthy environment. As the action plan concludes, "the long-term success of the more important initiatives such as the internal market and economic and monetary union will be dependent upon the sustainability of the policies pursued in the fields of industry, energy, transport, agriculture and regional development; but each of these policies, whether viewed separately or as it interfaces with others, is dependent on the carrying capacity of the environment."[14] This is the best definition of environmental limits I have seen.

In general, the European Union's action plan shows a high level of sophistication in its thinking about competitiveness and environmental quality, and indicates that the community's leaders have a clear grasp of the connection between the two. "Turning environmental concern into competitive advantage is one of the objectives of 'Towards Sustainability,'" it asserts. "By aiming at reduction and elimination of pollution and at prevention, recycling and reuse of waste rather than just abatement or clean-up and by creating a broader mix of instruments, including market incentives, thereby avoiding constraints on the technologies used to achieve higher standards, environment policy can stimulate investment, innovation and competitiveness rather than stifle them."[15]

Transforming economic patterns in the European Union so that sustainability can be achieved, the plan states, implies the recognition of at least three things: that economic and social development depend on the quality of the environment and on the satisfactory guardianship of its natural resources; that the reservoir of raw materials is finite, and the stages of processing, consumption, and use should be managed to aid and encourage optimum reuse and recycling, thus avoiding waste; and finally, that citizens' behavior should reflect an appreciation that natural resources are finite, and one's individual consumption of these resources must not be at the expense of another's, or of a future generation's.[16]

Structurally, the European Union's program is similar to the Netherlands' NEPP. It identifies priority environmental themes for the European Union: climate change, acidification and air quality, protection of nature and biodiversity, management of water resources, the urban environment, coastal zones, and waste management. Adopting the Netherlands' target group approach, the plan identifies industry, energy, transportation, agriculture, and tourism as the

European Union's key target groups, and has a carefully thought-out strategy and action plan for each.

However, *Towards Sustainability* is a guideline for European Union environmental policy, not legislation, and thus tends to be a much more general document than the NEPP. Its primary focus is on strategies and mechanisms rather than on quantitative targets. For example, strategies regarding the energy sector include the use of economic and fiscal instruments to encourage sustainable energy use, negotiating agreements with industry on efficiency, and the implementation of national energy efficiency programs that set efficiency standards, among other things. These process-oriented strategies set the stage for the quantitative, measurable goals that will be worked out in the future. However, some sections of the plan already have quantitative targets attached, such as the program to combat acidification, which calls for a 30 percent reduction of nitrogen oxides and volatile organic compounds and a 35 percent reduction of sulfur oxides by the year 2000.[17]

Other sections of the action plan deal with the management of risks and accidents and broadening the range of instruments, which includes discussions of the roles of research and technology, spatial planning, market mechanisms, education, and financial interventions such as subsidies and investment policies. Two sections are devoted to implementation and enforcement, which sectors within the community will be responsible for implementing which policies, and how the plan is to be integrated into the community's legal and political structures.

Although the European Union was created initially as a trade mechanism, it has learned that issues of the environment and of quality of life are equally important aspects of trade relations between nations. This sensibility, and a refreshingly ethical perspective, are expressed in such statements as: "there should be a clear understanding that certain aspects of the environment are or can be 'priceless' and thereby not susceptible to normal economic costing mechanisms such as cost/benefit analysis or the free play of the market forces, e.g. an adequate quality level of drinking water, the last giant panda or elephant, the singing of birds, aspects of cultural heritage."[18]

• • •

Towards Sustainability is a valuable example for a world that is hungry for more flexible trade relations, a standard of excellence in terms of combining economics and environmental quality. To understand its value, one need only compare it to the hodgepodge of so-called environmental provisions in such agreements as NAFTA and GATT, which are barely comprehensible and consequently will be a source of frustration and conflict in the future.

Because the green plan idea is so new, each of the countries discussed in this chapter – and each of the green plan countries, as well – have evolved their policies independently, with little sharing of knowledge and experience with others. In the future, countries evolving toward green planning will have the advantage of the pioneering experiences of these nations; they will have a much better idea of what works and what does not.

PART THREE

Identifying Ingredients for Success

8

Broadening the Scope of Resource Management: Principles and Techniques

The information presented in the previous section illustrates that green plans are far from the stereotype of rigid, monolithic, top-down government programs. There is no single green plan model that can be applied to every country and every circumstance, nor should there be. In addition, each country's green plan will change over time, as it responds to new information and to changing conditions and concerns. At best this book takes a quick picture of green plans as they exist now, but the process itself can really only be seen in longer segments of at least ten years. Only by studying it within this longer time frame can we begin to judge which aspects are successful, understand why others are not, and start to link the information from various countries' experiences.

However, the existing green plan efforts already offer important information from which any nation, industry, or household can learn. The ingredients for success discussed in the following section have been reviewed and tested by many dedicated and well-informed people who agree upon their usefulness. Any nation in the process of creating a green plan will want to study the principles and practices green plans have in common, because they are keys to success.

A green plan is not a project, but a process, one that involves a shift in thinking about the ways in which we interact with the environment. It means applying new concepts and tools to the field of resource management, and also many that are not new, that have been applied by businesses and other professions for years, but rarely by resource managers. Some of these elements are already being integrated into environmental policies in the United

States and elsewhere, but green plans are unique in that they pull together much of the advanced thinking on policy, management, and government-society interactions into one comprehensive strategy.

This chapter covers some of the ways in which green plans are broadening the scope of resource management to include other disciplines, and includes a few of the most commonly applied concepts and instruments. Chapter 9 discusses how the shift in management thinking in the green plan countries has led to a new working relationship between government and business to achieve environmental goals. Chapter 10 looks at the new relationship green plans develop between government and other sectors, from community groups to environmentalists to the individual, and how these relationships build the political and social base necessary for green plan implementation. Chapter 11 shows how all of these elements relate to the situation in the United States, and recommends strategies for green plan development here.

Multidisciplinary Resource Management

One of the key principles shared by all green plans to date has been to broaden the scope of traditional resource management to include social and economic factors. Up until recently, most of the research into environmental issues, and most of the thinking about resource management, has focused on the natural sciences, which have become more and more specialized over the years. Consequently, we know a great deal about certain parts of the picture, such as wildlife management or forestry or soils or air quality.

What we do not know much about is the interaction between social systems and natural ones, and how we can change the former to preserve the latter. We are handicapped because we do not have that information; if we are going to solve environmental problems, we need to know as much as possible about their causes. We may have made the problem of environmental decline worse by relying too heavily on the natural sciences in setting policy, when the social sciences, management concepts, and ethical considerations must also play a role.

A good green plan manager is like an orchestra conductor, pulling together all the disparate elements of the orchestra into one big team working toward the same goal. In this orchestra, scientists might be the violin section, environmentalists the cellos, economists the bass, business leaders the woodwinds. The orchestra would not exist without these sections, but it is up to the conductor to pull them all together, the music of each section heard and yet smoothly woven into the whole.

Making our way toward environmental recovery and sustainability involves a great deal of uncertainty. Our knowledge about both natural and human systems is still limited, and what we now believe we know will not prove to be entirely right in the long run. We tend to believe that the natural sciences can provide us with the answer to a problem, but science rarely deals in such certainties. We need to be prepared to evaluate new evidence, to change our minds, and to use our own best judgment. This kind of flexibility is built into process-oriented approaches such as green plans.

The Dutch and Canadians, among others, have started shifting some of their emphasis to research in the social sciences. It is very important that the United States begin to understand the lessons these other nations have learned and to focus our research accordingly. We need to have sociologists, economists, psychologists, management specialists, and others involved in defining how we are going to manage our natural resources into the future.

Integrating Environment and Economics

The social system with the greatest impact on environmental quality is, of course, our economic system, with its subsystems of industry, agriculture, trade, and so forth. That is why some of the most significant forays into the social sciences by innovative environmental policy makers have been in the area of economics. Although much work remains to be done in this field, there have been important advances in understanding how and why economics affects the environment, and in developing policies to counteract negative effects and encourage positive ones.

A healthy environment is essential to a healthy economy, yet environmental considerations are not given the same weight as economic ones in most of the decisions we make. Despite the fact that we depend on the natural environment for our quality of life – for our very existence – we continue to degrade and deplete such natural "commodities" as air, water, soil, minerals, and energy sources as if there were no cost involved. We have treated these resources as free, even though their actual value to us is very great. We can no longer afford to ignore the costs of environmental degradation in our economic practices, nor can we continue to displace those costs onto other countries or future generations.

If we do continue in our current pattern, environmental problems will increasingly cause our economy to suffer as well. As one report from the Netherlandic government says: "We used to look with concern at the detrimental effects of economic growth on the environment. Now it is high time to concern ourselves with the disastrous consequences for the economy of environmental destruction."[1] Environmental concerns must become an integral part of all economic planning, whether on the macro or micro level.

On the level of government policy making, green plan countries encourage this integration by making their plans the product of a number of important government ministries, not just the environmental ministry. The result is that the authorities responsible for trade and transport and agriculture learn to consider the environmental consequences in all the decisions they make and policies they implement.

To integrate environmental concerns into *all* levels of economic decision making, green planners around the world (and other environmental researchers as well), have been studying ways of modifying economic theories and models to include environmental "costs" that the current methods ignore. One example is the concept we call GNP, which was developed during World War II to measure the scale of the war effort. It is clearly limited as a method for measuring what is happening in the real world today. Developing an equivalent that factors in environmental data is proving very difficult, but the Norwegians and the Dutch have done some important preliminary work on natural resource accounting. The Dutch in particular have done very sophisticated studies on environmental indicators. Like economic indicators, environmental ones are used to measure the health of larger systems, and they can be used to measure impacts, or costs, of certain behaviors on certain resources.

Green planners are also experimenting with various ways of incorporating environmental costs into our pricing system, influencing everyday decisions in the marketplace. Market-based mechanisms cannot replace such tools as regulations, ambient standards, and permit requirements, but they do have advantages that the traditional environmental controls do not. They are typically much more flexible than regulations, and usually have a more direct impact on economic decision making. They can be very effective in getting businesses and individuals to voluntarily move in the desired direction. They include things like taxes and levies on "bad" environmental behavior and financial incentives for "good" behavior, and abolishing environmentally harmful subsidies.

Currently, one of the most frequently applied market-based tools is the polluter pays principle, which in essence means that the user of a resource pays for the negative effects of that use. For instance, the price of a car does not include a whole array of costs that society must pay in order to allow you to operate it, such as road improvements, the building of freeways, traffic police, licensing procedures, and snow removal equipment. If these costs were included, the price of owning the car would be much higher. The same is true in terms of costs to the environment: take, for instance, air pollution from traffic, which remains one of the last unsolved air quality problems for most cities. These extra costs have been coming out of the taxpayers' pockets, including those of people who have never driven a car in their lives.

A number of countries, including most of the European ones, began to apply the polluter pays principle to gasoline consumption years ago, which is one of the reasons why some have a gas tax that is more than twice as high as it is in the United States. Rapid transit, parking, and road and bridge building are all paid for by the auto drivers; that way, a non-auto-user is not penalized by having his or her tax dollars used to clean up all the problems caused by cars.

Cutting harmful subsidies is another effective market-based tool. In the past, governments have far too often subsidized activities that were environmentally unsound. For instance, the government of New Zealand used to subsidize the clearing of marginal farm land, most of it in steep, forested regions. Because it was so steep, the land was not capable of sustaining an economic crop, and a great deal of soil erosion resulted. In one region, the government had to pay to replant trees it had previously paid people to clear, in order to stabilize the erosion problem. The effects of abolishing these types of subsidies can be dramatic, as New Zealand discovered when it got rid of all its agricultural subsidies. Farm use of chemical pesticides and fertilizers has dropped substantially, because without the subsidies they no longer make financial sense.

Conversely, the Netherlandic government uses some subsidies to encourage behavior that it wants, but has hesitated to rely on them because they are vulnerable to budget cuts. It prefers to work with the polluter pays principle. For example, it levies taxes on households for waste pickup, then uses those funds to subsidize recycling programs. The Dutch like to think in terms of cycles, so they tend to look at things like "closed money streams," in which they might use extra taxes on fuel-inefficient cars to subsidize rebates for more fuel-efficient cars, or use taxes on pesticides to subsidize more sustainable agricultural practices.

Other countries, like Sweden, have also had good results using this type of mechanism. The Swedes finance 90 percent of their environmental program through taxes and levies applied on the polluter pays principle. They have a whole battery of environmental "sin" taxes: charges on carbon dioxide, nitrogen oxides, and sulfur that are designed to reduce emissions to certain target levels; on pesticides and fertilizers; on mercury and cadmium in batteries; and heavy taxes on oil.[2]

Environmental taxes can provide a great deal of financing for environmental programs. However, their main purpose is as policy levers, to influence the behavior of consumers and industry.

Tradable pollution permits are another market-based incentive that have become quite popular of late, but because I believe it is critical to focus on pollution sources, I am not enthusiastic about the idea. Even if they make sense, as presumably they would in some instances as a way to lessen the costs of regulation and therefore make it more palatable, the politics involved are devastating. They may create the notion of a "right" to pollute; they will certainly create an economic value for pollution, when it is really a cost. A Wall Street-based permit trading market will come complete with special interests that will ensure that the policy is never changed. Although the popularity of these permits is assured, the claim that they will have positive long-term effects on the environment is debatable.

Another important cost-related principle that green plans incorporate is pollution prevention, or using source-oriented measures rather than effects-oriented measures. Because the long-term costs to the environment are frequently not considered when a decision is made, current policies, governmental and nongovernmental, are largely limited to putting out fires that have already started. It makes a lot more sense, both environmentally and financially, to prevent them in the first place.

Industry has learned repeatedly in this century that it always costs more to clean up afterward. If, for instance, a company pollutes a river and then has to clean the riverbed for a thousand miles because there are heavy metals in it, the cost is staggering – far greater than it would have been to find a cleaner way to operate. The same is true with pollution of aquifers. In the early days, a lot of companies just pumped pollutants into the ground, where they seeped into the aquifer, spreading in an underground plume. This is extremely costly to clean up, and sometimes is simply impossible to do.

One way to avoid this problem is to change production cycles to stop pollution and waste, rather than simply improving "end-of-pipeline" technologies. Many innovative examples of this approach exist, some of which were being implemented in industry years ago. One such example is that of the Dow Chemical plant in Pittsburg, California. Twenty years ago the plant was releasing polluted water into a river that ran alongside it, and was also using a huge amount of water in its manufacturing process.[3]

Pollution laws pushed Dow to find ways to treat the contaminated water before release. Its first solution was to install solar evaporation ponds, but that process created a toxic sludge that had to be disposed of every few years. In the 1980s, California law changed and companies were forced to abandon solar evaporation ponds. Dow decided to face this new challenge head on, initiating a major research effort called "Off the Ponds" to develop an alternative. Eventually the company found a way to treat the water for reuse within the plant.

As a result, Dow significantly cut its water use, recycling an average of 150 gallons per minute through this water reclamation system. The company decreased its costs because it used recycled water rather than buying additional water to pump in. Most significantly, it saved millions of dollars it would otherwise have had to spend constructing a new water treatment plant.

The Netherlands' NEPP places a great deal of emphasis on this sort of change in production patterns, but goes even further, encouraging increased efficiency throughout the entire life cycle of a product. This has stimulated such innovations as the car recycling effort by Volkswagen discussed in chapter 4. The NEPP also aims to change consumption patterns in that country, using incentives, information and education campaigns, and regulations. They call this "external integration"; businesses and consumers learn to integrate environmental concerns into all their choices and practices.

Making Complexity Manageable

Another management principle adopted by green planners is that it is important to be able to understand and articulate policy goals on an extremely broad level. The big picture, systemic approach to environmental policy is necessarily quite complex, and its complexity makes it very difficult to deal with, scientifically, managerially, and politically. Given that the human mind can only process a certain amount of information at a time, how does one get a grasp on the complex tangle of natural systems and human affairs and policies and projects, all of which need to be considered? How does one communicate a very complex idea to political leaders or to the general public in terms they can understand?

The Dutch developed an answer that represents a major breakthrough in environmental planning: the idea of folding all environmental issues into an overarching framework of eight themes, or issue areas. This breakthrough might not have been possible had the Netherlands' minister of environment at the time, Pieter Winsemius, not been a management specialist. Outside of government service, Winsemius works for one of the world's best-known management consulting firms, McKinsey & Company. He earned a doctorate in physics before completing an MBA at Stanford University. Winsemius had the knowledge and experience to apply good management practices rarely seen in resource management agencies, whether in the Netherlands, the United States, or anywhere else.

Those skills did exist within these agencies, but previous leaders did not have Winsemius' experience or vision. His ability to maintain his management perspective in the face of the competing interests and the daunting complexity of the environmental dilemma, combined with the technical skills of the very capable staff in his department, has been one of the keys to the NEPP's success.

When Winsemius took office, the environment ministry staff told him that the Netherlands could set an environmental example for the world if it could find solutions to about thirty separate problems. One of their sources of inspiration for this idea, the U.S. Environmental Protection Agency (EPA), had more than fifty priorities *it* was trying to solve. Winsemius knew that the human mind can only grasp a very limited number of concepts at one time, and that it would not be possible to actually manage this number of priority problems. Under his guidance, the environmental ministry cut its priorities, combining individual issues into five generalized types of problems, or what they term "themes." (Three more themes were later added.)

The themes they settled on were: climate change, acidification, eutrophication (the disruption of ecological processes caused by an excess of nutrients in the environment), the diffusion of toxic substances, waste disposal, disturbance (noise, odor), dehydration, and squandering. They did not leave any important issues out; they simply subsumed them within the themes. For each different theme, they listed a series of actions to be taken to correct the problem.

This one element completely altered the viewpoint from which the Dutch approached the problem. It fit all the complex ideas and their many ramifications into a definable, understandable package. Most important, it made it all manageable: goals could now be developed on every level, from the most broad to the most specific, and progress toward (or away from) those goals could be measured and reported in ways that everyone could understand. Even though I have had an entire career in the field, I had never found a satisfactory way to get beyond that intricate tangle of factors and issues involved in resource management, until Winsemius showed the way.

Using Information and Technology

One of the most important tools for dealing with the complexity of environmental issues and planning is an information system that includes a centralized resource database. With such a system, the specialists working on various elements of the program can always relate their part of the project back to the central idea. The United States suffers greatly from the lack of such a database; until one is created, it will be very difficult to move ahead with any sort of comprehensive environmental planning.

One of the reasons that the NEPP is such a strong program is that the Netherlands has a central information database. Because of it, the Dutch are able to tell what the real environmental conditions are at any one time, so that an agency or department can see the results of its policies. A comprehensive program can only be effective if you have goals and can demonstrate progress in this fashion.

The Dutch have also become very skilled at computer modeling, using the database information to create a better picture of the way in which natural systems work. It is the computer's ability to show the relationships across a broad spectrum that allows us to move into the new realm of comprehensive thought and planning; computers make integrated management possible. Until they were developed, it was much more difficult to see and understand the interrelationships between resources, within ecosystems, among humans and the environment, and so on. A good computer model also has the advantage of being able to calculate instantly the outcomes of various approaches, allowing decision makers to weigh the advantages and disadvantages, risks and benefits, of alternative policies.

There are endless examples of how dramatically the computer has changed things. For instance, as a student many years ago I was intrigued by the idea of using overlays in studying land-use problems like erosion in a watershed. One would start with a basic map of an area, then add clear plastic overlays with forest types, soil types, altitudes, and a half-dozen other elements. Overlays tremendously enhanced the understanding of the problem and the decision-making process; at the time, they were an exciting breakthrough. It was difficult, however, to use more than a few at a time, since they had to be actually tacked onto the map.

Now, with computerized geographical information systems (GIS), it is possible to apply hundreds of overlays at a time to a computerized map. At the touch of a button, you can overlay soils in one color, altitude change in another, wetlands, pollution sites, densities, highways, and on and on. Hit another button and you can see the potential effects of an oil spill—where the river would carry the oil, for example. GIS is a rapidly developing tool, unavailable just a few years ago, and is an example of what the immediate future is going to look like in terms of resource management.

Although the computer is an important tool, it is good to remember that it is just a tool, not a solution. A computer is only as good as the information with which it is programmed, and a computer model is only as good as its creator. That said, one of the most significant things about the computer is its potential to make a vast array of information available to everyone. If a huge data library with information about resource issues were available to the public, any school child doing a research paper or any business manager or public agency manager would have at his or her fingertips a tremendous new source of knowledge.

This access to information adds tremendous strength to traditional advocacy. It provides a way to better understand and intelligently consider issues that have become clouded by emotional undertones. For example, any intelligent user will instantly wonder how many people the earth can hold and still have enough water, clean air, and other essential resources. Such a data library can quietly show the need for population policies – which have been very difficult to achieve – because the data on population and resources speaks for itself.

Computers can also allow us to make sophisticated use of economic information as it relates to environmental planning. Funding should be provided for fiscal specialists to gather the information on the costs of programs and improvements. A database can help us assess the real costs of various human activities – for example, what automobile use costs us in terms of air pollution, road building and maintenance, energy consumption, and so on.

No entity in the world is going to pull together *all* the information that could be useful, but collecting whatever information is available regarding a range of issues, be they global, national, or local, will make decision making a hundred times more meaningful. We need not be utterly precise with this data bank and the questions we ask it; because we have pulled in the best information we have at hand, the basis for our decisions cannot help but be much improved. In the past we have had to rely far too much on blind guessing; now our guesses will at least be more informed.

Goals, Timelines, Monitoring, and Reporting

Another good business management practice green plans make use of is goal-oriented planning – management by objective. The first step is to identify problems and agree on objectives in the problem areas, such as combating ozone depletion by phasing out CFCs. Identifying environmental objectives should be done in consultation with all interested parties – environmentalists, business leaders, everyone who is concerned.

Once objectives are identified, planners can move ahead with setting more specific goals and the timelines for them. For example, one of the Netherlands' goals is to reduce its emissions of volatile organic compounds 60 percent by the year 2000 from 1981 levels.[4] Goals like this should be based on the best possible science, and the time constraints should be realistic and based on the risks involved. In the case of CFCs, for instance, the risk was so high that it was necessary to phase them out fairly quickly.

It is very important to lay out specifics such as these as clearly, and as early in the process, as possible. The interested parties, particularly in business, must be able to see and understand where the plan is going, and they must be given time to develop strategies. Defining and clearly stating the goals may be the most crucial leadership role for government in the process.

Specific goals and timelines are a way to measure progress toward the solution of a particular problem. Once you establish where you want to go and how fast you want to get there, you can monitor conditions and keep measuring what is really happening against those standards. The original goals and timelines may have to be changed if new information comes to light or if indicators show that they are not working to solve the problem, but this kind of flexibility needs to be built into all green plans. Monitoring also allows you to respond quickly to any unforeseen effects a policy might have and any new problems that may crop up.

Regular monitoring is thus another key management tool used in green planning, as is reporting the results to the public. The public needs to know where things stand – what progress has been made, what things have not worked. If people are well-informed, they will be able to make better decisions. They will also have more faith in the seriousness of the effort, and building public trust is essential to building public support for the plan.

Both the Netherlands and New Zealand, in addition to having their own governmental procedures for assessing their programs, have asked independent entities to monitor the progress of their green plans, as an outside control. In the Netherlands, the National Institute of Public Health and Environmental Protection, a highly respected independent research institute that wrote the initial report *Concern for Tomorrow*, will update the NEPP's information every two years. Like a state of the environment report, the update looks at how well the plan is being implemented and enforced, how effective its programs are, what the contributions of different measures are to the total solving of the problem, and so on. Every fourth year the institute will complete a forecasting report, looking at whether or not the plan is still on track, whether new themes should be added, and what new problems are emerging that will need to be coped with.

These reports are detailed and extremely honest. The Dutch seemingly do not hold back anything from their own people or from the rest of the world. This honesty is a basic ingredient to maintaining public trust in a program. One reason the NEPP has succeeded where most traditional planning programs have failed is that the government does not try to fool those who are involved in the process, especially the public.

New Zealand established the Office of Parliamentary Commissioner for the Environment as an independent authority to review and publicly report on the environmental effects of government policies and programs. The commissioner's office reviews the government's resource management system, determines its effectiveness, and investigates incidents of actual or potential harm to the environment.

The commissioner will also conduct inquiries and provide reports on a variety of matters to the House of Representatives. Both this office and the country's green plan are very new, so the ways in which the system will operate have not yet been completely worked out, but it already seems clear that the Office of the Parliamentary Commissioner will be an important player in the process.

Bringing together all of these solid management principles and techniques under one green plan umbrella is a truly remarkable accomplishment, one that would not have been possible without the dedication and talent of the government officials who wrote, worked for, and are implementing their countries' green plans. Government staffers are rarely credited for what they do, but these men and women really were remarkable pioneers in this field, and their contributions should be noted.

One of these is Paul de Jongh, who at the time the NEPP was written was a junior staff member in the Netherlands' environmental ministry. He was so well-informed and dedicated that he became the primary source for much of the environmental information in the NEPP, as well as a constant critical voice in the process. When the original version of the plan bogged down during the writing phase, de Jongh was brought in to write portions of it. Today he is a deputy in the ministry, and responsible, along with two other individuals, for the NEPP's implementation. There were many others, of course.

The senior civil servants in New Zealand's government also provided some very courageous leadership, because what they set out to do was totally restructure government's approach to management. They created new agencies, eliminated many old ones, and established different lines of responsibility. They completely changed the way things were managed, even though in some cases that meant eliminating their own jobs. It was a remarkable test of the integrity and professional commitment of a civil service.

In terms of leadership, the senior civil service sector also does some very important promoting of the green plan idea, both in their own countries and around the world. It was a focused group of these individuals – Robert Slater, who headed up Canada's green planning process from its beginning, Paul de Jongh from the Netherlands, Paul Hofseth from Norway, and others – that worked quietly in Rio to make sure that the idea of comprehensive national environmental strategies was woven into the fabric of Agenda 21. Because of these people and others like them, Agenda 21 will have a tremendous impact in the next century.

9

A New Relationship between
Government and Business

One of the powerful advantages of the comprehensive approach to environmental policy is that it includes more than the typical legal, fiscal, and technical elements: it also involves a breakthrough in cooperation. For too long there has been an adversarial relationship between business, government, and citizens, and everyone has suffered as a result. The truth is that we need each other. Green plans offer the opportunity to learn a new way of working together toward environmental health.

We cannot make real the dream of solving our environmental problems without the active cooperation of business, which may be the single most important factor in the success of green plans. Sustainability cannot be achieved without it.

It is also very much in the interest of business to join in the process. First of all, a healthy economy depends on a healthy environment. Second, green plans are the future, and businesses that do not join in will fall behind. Third, green plans offer businesses the chance to escape the frustration of regulatory chaos in exchange for their cooperation.

Time for a New Dynamic

The business sector has tremendous influence on the policies of western democracies. In the past, it has used this influence to oppose environmental planning and legislation it perceived to be against its interest. But just as business has been a very powerful political opponent, it can be a powerful ally. We have a choice: we can keep fighting political and legal battles, or we can start to build relationships among business, government, and environmentalists to solve the problem. Some businesses already understand, and others can be brought to see, that planning for sustainability is to their advantage in the long run.

Because industry is a principal source of environmental problems, it needs to become a principal source of solutions. Businesses that have caused problems can be the most effective at fixing them, since it is likely that they are better equipped than government regulators to understand the often complex technology involved in their manufacturing processes. It is usually costly and inefficient for government to try to micromanage business practices.

If we can find a way to make it easier for the business community to participate on a voluntary basis, we will be much more successful much more quickly. This is one lesson the United States can learn from Europe: there is no real reason that government and business cannot work together toward environmental goals. During the last few decades, U.S. government officials have been opposed to what they viewed as government interference in industry, but in fact government and business have never been and cannot be entirely independent of one another. Even now the U.S. government provides subsidies to businesses and regulates some of their activities, yet it rarely does so with any clear vision of long-term goals that will benefit both business and society. It has become quite clear that such a vision is needed to promote government-business cooperation toward cleaner industrial processes.

Closer working relationships between government and business are not completely unknown in the United States. For example, during World War II the government had to set industrial priorities, such as requiring certain industries to shift to weapons production. It also had to regulate the petroleum industry, because we could not continue to use unlimited amounts of gas for automobiles if we were to provide fuel for the military effort.

Traditionally, however, relationships between industry and government in Japan and many of the European countries have been far closer than in the United States. Every major Netherlandic, German, or French industry receives some assistance from government, particularly when it comes to funding research operations. If a new product or technology is feasible for the long term and gives the company the opportunity to enter new markets, the government usually helps to develop its potential.

In return, the government gets a healthier economy and increased tax revenues. It also gains the advantage of a good working relationship with industry in which both parties can sit down and negotiate plans for achieving environmental goals. Even though they may have different viewpoints, they are able to discuss and compromise and come to an agreement.

This is the sort of relationship that nations around the world, particularly the United States, need to develop. The time is right, because industry's attitudes toward the environment have been changing. Over the last few years, more and more executives have come to understand that environmental concerns must become an integral part of their company's decisions.

One reason is simple competition. Businesses in Japan, Germany, and other European nations are switching to cleaner manufacturing processes in order to avoid the costs of cleaning up after the fact – one of the principles of green plans that is often far more efficient and cost-effective. Competition in this area has awakened many industrial leaders.

But another reason is that there are an increasing number of sophisticated companies managed by people who see themselves as citizens as well as business leaders. They realize that their behavior helps shape the kind of world in which their children will grow up. Edgar S. Woolard, chair of Dupont, is one example. Woolard was quoted in an editorial published in the *New York Times* on Earth Day 1990, saying that henceforth Dupont would be an environmental company, because the American public would no longer accept the practices that had led to environmental decline in the past. To the company's competitors who had no plans to change, he said that Dupont looked forward to taking away their business.

This is happening more and more in the business world; other examples of corporate leadership include 3M and Dow Chemical. Dow used to be a particularly tough opponent of environmental regulations. At one point, the story goes, the residents of Dow's "company town," Midland, Michigan, began to feel unhappy about the company's reputation as an enemy of the environment. They wanted a clean community and a clean state; they recycled and planted trees and conserved energy. But they realized that Dow's reputation reflected badly on them, and people thought of Midland as a nasty, dirty place to live.

Some of the community's churches and civic organizations and individuals began to pressure Dow to figure out how to manage its environmental problems instead of fighting regulation. They thought that if the company put its resources and energy to work on better environmental practices, it could turn its reputation, and Midland's, around. Dow listened, and what it heard began to affect its policies and plans. Now the company is becoming known as a national leader among environmentally progressive businesses.

The Netherlands' experience has shown it is wrong to think that businesses are always environmental "bad guys." Once a business decides to become a willing partner in environmental recovery it can often do more than the average citizen.

My experience in this area can be summarized by a talk I had with the owner of a shipyard in the Netherlands that builds some of the world's finest sailboats. As we stood before his brand-new, environmentally innovative facility, he said: "You know, I built this three years ahead of the required deadlines. I felt that it was the ethical thing to do, but of course I had some financial help from the government."

Then he took me around and proudly showed me the various innovations, which included a system that completely changed the air every six minutes, because of the fumes from paints and other toxics. In order to do that he had bought the kind of air purifying system used on large jets. As we looked at all these innovations, I realized that he was indeed part of a national purpose, and that he knew that what he had done would help not only his own business, but would encourage others to do the same. His was not an isolated case; his attitude is shared by a great many of the Netherlandic business people I have met.

It may take some time for a business to come around to this attitude, but there are ways that government can speed the process along. A good example is the controversy surrounding energy in California in the 1970s, when I was secretary for resources. The state government then was able to change the attitudes and directions of the powerful energy-producing industry by using both regulation and incentive.

The industry was determined that it had to keep increasing its generating capacity; its plans included forty nuclear power plants. The resources agency was equally determined that a combination of conservation and alternatives could provide the answer. The debate was long and heated, but eventually everyone realized that the efficiency obtained from better management of the existing supply of energy would save a tremendous amount in money and resources, and would be a far better strategy than worrying about creating more.

Once the utilities saw that it was in their own interest, they took off on their own initiative and became innovators in the field of conservation. Other private businesses, such as banks and chemical companies, followed suit. All of this happened because, in addition to telling the utilities they could no longer operate in the old way, government gave them an incentive to change by making it more profitable to conserve energy than to create new sources. That is a model that the rest of the world is attempting to duplicate. We have to do the same thing with the oil industry – make conservation more profitable than increased production – in order to win its cooperation.

Eliminating the Frustration Factor

Dealing with overlapping and outdated regulations is one of the biggest frustrations industry faces today, which is why green plans have been so successful in attracting business's cooperation: they make the streamlining of environmental regulations and requirements a priority. They also clearly state the government's policy and lay out the long-term environmental goals; business is not kept guessing about what the government wants, and does not have to worry that policies will keep changing from year to year.

The problem with most countries' regulatory systems is that they have been patched together one regulation at a time. Too often, new regulations were passed without much thought for ones that came before. Some became obsolete but were never abandoned. The result is a kind of regulatory chaos: laws contradict each other, they overlap, they require mountains of paperwork and outdated compliance measures. Only lawyers really understand what is going on. Every day a company faces the prospect that a regulator will walk through the door, demanding a new permit or another report. It is a tangled maze that has grown thicker and thicker over the years.

An everyday example of this sort of thing is that of a woman who owns a bus service in California. She recently tried to build a site for passenger pickup, and found that she needed thirty-nine permits from different departments. This process eventually cost her $100,000 in legal fees and technical expenses. After all that, the regional council voted down the plan because five neighbors did not want the bus stopping in their neighborhood.

A dedicated environmentalist who was trying to promote efficient mass transportation, this person ran into an endless web of bureaucracy that in the end proved unworkable for her. After risking her own limited funds, she was forced to give up the project. Thousands of automobiles will remain on the roads of that community because there is no mass transit alternative.

This type of regulatory nightmare is costly for both business and government.

It is inefficient, and often does not help to achieve environmental goals. If, as is being demonstrated elsewhere, cooperative action really works, then getting rid of this accumulated frustration may be the only incentive business needs to get out in front – to not only cooperate, but lead.

In the green plan countries, government's ability to solve the frustration factor was a major reason businesses chose to cooperate. These countries made it a priority to revise or eliminate overlapping and conflicting environmental regulations; in return, the business sector has accepted the principle of sustainability.

Both the New Zealanders and the Dutch have made particularly strong efforts to find new ways of working with industry. In both countries, the government sets environmental goals and provides businesses with the policy framework they need to achieve those goals. Within that framework, businesses are free to determine how they will meet the goals. Financial incentives are used to encourage environmentally sound practices, and regulations are kept in reserve to make sure that no one cheats.

Most important to the idea of sustainability, these governments encourage their industries to look at how pollution and energy waste can be eliminated at the source. This puts business on a different course, one in which environmental concerns are integrated into all of a company's planning, playing an important role from the drawing board to the finished product and beyond, to the way in which the product is used and disposed of. In the long run, integrated solutions prepared by industry are the best way to achieve sustainability.

It all boils down to a new way of looking at how government, business, and society interact to solve problems, one that stresses negotiation, incentives, and good management over confrontation and regulation. The Dutch even go beyond the idea of compromise and aim for consensus in their policy making, because they believe it strengthens the policy's acceptability and thus makes it easier to implement. If all the interests involved agree on a policy, they do not have to worry about one or two being able to block it, as the oil and coal interests blocked the Clinton administration from setting a new energy policy in the United States.

This new relationship between government, business, and society is really about stewardship, an agreement that we will take better care of the earth – all of us, including the business sector. This requires a transition from development without limits to an acceptance of limits; to an era of thoughtful

141

conservation and movement toward sustainability.

Such a transition will inevitably involve careful, long-term planning, but businesses are well acquainted with planning. Most companies prepare plans on a regular basis, laying out the actions that will be necessary in order to reach manufacturing, marketing, and distribution goals, and including an estimation of costs and an evaluation of potential efficiency improvements. This is exactly the sort of evaluation that green plans encourage, except that "green business plans" will make environmental goals an integral part of all the company's decisions.

The Netherlands' Target Group Approach

The best example of how effective a green plan can be in helping government and business work together is the Netherlands' NEPP. (While New Zealand's plan also contains a dramatic new approach to the government-business relationship, it has not been in force long enough to determine how effective it will be.) The Dutch have moved quickly to establish this new relationship, and have achieved some tremendous results with their target group approach.[1] The United States would do well to adopt a system based on this model.

The Netherlandic government realized that asking business to integrate environmental concerns into all its planning and processes would mean asking it to develop completely new ways of thinking and operating. It also knew that, in order to achieve such a fundamental change, it would have to work closely with industry and consumers. From the very earliest stages of the NEPP planning process, the government reached out to business representatives, to get them involved in the process and to solicit their willing participation in the plan.

And business did get involved. When the NEPP goals were published, industry representatives came to the government and said that they were willing to commit themselves, even though the requirements were very strict. In return, they wanted a commitment from government to allow them to work out their own solutions to the problems. They also wanted the government's policy to be consistent and long-term, so they would not have to worry about constant changes in the rules.

Netherlandic businesses act in their own interest, as do businesses everywhere. Like their counterparts in the United States and other countries, industries in the Netherlands had been largely resistant to environmental initiatives

Soil Cleanup in the Netherlands. This soil cleanup operation in the city of Utrecht is just one of many projects aimed at reducing the Netherlands' high levels of soil contamination. One of the first initiatives the government undertook in partnership with business was to list all sites where the soil had been contaminated and develop cleanup plans. Photo courtesy of the Netherlands' Ministry of Housing, Physical Planning and the Environment.

prior to the NEPP, fighting more stringent regulation and minimizing the costs wherever possible. They agreed to the NEPP goals, even though it would decrease their profits in the short term, because they had come to realize that it was in their interest to have a clean environment. If there is no environment, they realized, there will also be no economy. Since then, Eastern Europe has provided them with some very sobering evidence that they made the right decision.

The business community also realized that by cooperating it could ask to be a part of the process and help to determine its own future, rather than waiting for regulations to be handed down. A final factor prompting industry to cooperate was the increasing pressure from the public – banks, insurers, consumers, and others – in favor of greater environmental protection.

The NEPP planners managed to come up with a strategy that answered the needs of both government and industry. First they identified about fifteen sectors of industry that are responsible for 80 to 90 percent of environmental pollution. They asked these sectors to organize themselves into sector or target groups to be represented by one trade association, so that government could deal with the whole industry through one representative group. At the same time, the target group manager, an official from the environmental ministry responsible for that particular target group, pulls all the governmental efforts together on its side of the negotiating table.

The main business-oriented target groups (there are also target groups covering various other sectors of society) are agriculture, traffic and transport, industry, the energy sector and refineries, the construction sector, and environmental production firms (e.g., waste disposal companies and water supply companies). These target groups have been broken down even further for the purposes of negotiation, into such categories as the primary metals industry, the printing industry, and so forth.

After the target groups were established, scientists set overall goals to be achieved by each one, on a broad range of issues, within the twenty-five-year period of the NEPP. Each target group was asked to come up with its own answers as to how it was going to meet those goals.

The next step in the process, which has not yet been completed for all the target groups, is for government negotiators to sit down with representatives from each of the industry groups and negotiate a voluntary agreement, called a covenant. Working within the broad framework established by the NEPP scientists, the negotiators hammer out the specific goals for that industrial sector, as well as a timeline for accomplishing them.

This means dealing with a wide variety of issues, sometimes all at the same time. The whole target group wants to know the sum total of what the government is going to demand of it, so that each industry and company will be able to plan for it. There is a great deal of negotiating, not about the long-term goals, but about things like the speed with which they are going to be reached, what the intermediate goals are going to be, and whether or not the responsibility for certain decisions will lie with the companies, the associations, or the government. Once the agreement is approved, each company in that sector group must put together its own green plan.

The government took the risk of making these agreements voluntary because it wanted to establish a relationship of trust and cooperation, and it worked: it has been able to secure the active cooperation of most of the Netherlands' businesses. At the time of this writing, about thirty covenants had been signed, including several with major industrial sectors, and negotiations are proceeding on the rest. This is not to imply that industry always agrees, or cooperates, with government positions. No one expects businesses to become environmental organizations, or vice versa.

The voluntary nature of the covenant process does not eliminate the need for legal measures. Laws and regulations and permits still exist in the Netherlands, and always will. Industry itself understands the need for both the carrot and the stick in achieving environmental goals, because it is always difficult to focus on the long term rather than the present. In fact, the business sector has insisted on the need for regulation to ensure that all companies are forced to adhere to the same rules – they want the government to make sure no one cheats. However, in the end a great deal more can be accomplished through cooperation than through regulation.

To date, agreements regarding the reduction of emissions have been signed with the primary metals industry, the electroplating industry, the chemicals industry, the dairy industry, and the graphics industry; seven more are expected within the next two years. The government first approached the large industrial sectors; once agreements have been reached with all of these, it will shift its attention to the smaller industries and user groups. These groups will be able to learn from the experience of the large industries, which will be developing tools such as environmental management systems, company environmental plans, and procedures that the others will be able to adapt for their own use. If the initial agreements are successful, they will help secure the involvement of the rest of the country's industries.

Covenants are also being negotiated that go beyond emissions reductions. For example, there are a number relating to the environmental quality of products, the best-known of which relates to product packaging. Agreements regarding energy conservation have been reached with approximately twenty-five branches of industry.

The administrator who is responsible for implementing the target group program said that, if all goes as planned, his department should be shut down within ten years. By then, they will have reached agreements with each sector, and the oversight and enforcement will be in the hands of the communities in which the plants are located.

The scale of what the Dutch are undertaking needs to be appreciated. There are more than 12,000 factories involved in the NEPP's target group sectors, each employing more than five people. Creating a workable plan that involves all of them in the process of sustainability is no simple task, but the Dutch are on their way to doing just that.

Advantages to Government and Industry

Obviously, the scale and complexity of these changes pose a real challenge to industry. Why do the companies agree to cooperate? The Netherlands' success with industry is not simply a matter of making it more difficult and costly to pollute. The NEPP's target group approach offers a number of distinct advantages to businesses: they are given the chance to work out their own methods for solving the problems they cause; they are given a reasonable period of time in which to find and apply those solutions; and they are saved much needless frustration and expense because of the government's streamlined regulatory approach.

The long term perspective gives industry time to adjust. Major changes like this cannot be accomplished overnight, and the NEPP's twenty-five-year time frame gives the parties time to negotiate the short-term methods and gives the companies time to integrate necessary changes into their plans and operations. Each target group also negotiates a separate time frame for its particular agreement. One of the first agreements to be signed, with the primary metals industry, states that the sector must achieve its environmental goals in ten years, which is the period the industry asked for. This amount of time gives the individual companies flexibility regarding their other development and investment decisions, allowing them to fit whatever they do to the environmental concerns that have to be factored in.

Predictability is another major advantage industry gains from a long time span. After an agreement is signed, companies should not have to face any major policy changes for a long time, which is more stability than they would find in a typical regulatory environment. Businesses will concede a lot in return for this advantage.

The government has made stability a major theme in its appeal to businesses. As a brochure for foreign businesses put out by the Netherlands Foreign Investment Agency states: "Dutch environmental regulations are considered strict, and it is difficult to 'cut an easy deal.' Because of this, however, industry is well protected against increased regulatory pressure – as may be expected in other countries where regulations are more lax – and against future liability for pollution or industrial accidents." The article is entitled "Up front, you know the challenges; up front, you also know the rules."

The other major advantage the NEPP has for business is that it streamlines all of business's interactions with government. Each target group is assigned a target group manager who is responsible for everything in government as it relates to that group. In the old days, if farmers had a problem regarding pesticides, they would discuss it with the division of toxic substances; they would discuss a manure problem with someone else, a water quality problem with someone else, and so on. They would lobby the agriculture ministry regarding environmental regulations. Now their target group manager deals with every single item that affects the agricultural sector, whether it is noise reduction, pesticides, equipment, manure, or any other issue.

The permitting process for business has also been streamlined, so that it is effectively a one-permit system; each facility gets one permit that covers everything, from the building to emissions to health concerns. This simplification doesn't mean that standards are more lax, but that the process itself is more efficient.

All these efficiencies have been made possible because the NEPP set up a comprehensive approach at the cabinet level. The cabinet in California, and the United States in general, like the cabinet in the Netherlands, sets government policies in such areas as agriculture, business, health, and education. What the NEPP does is pull together four cabinet posts that are very critical to its function: agriculture, economics, environment, and traffic and transport. All four of these cabinet ministers are involved in the process, and all must approve the various policies and agreements. That is quite rare, and it provides a tremendous philosophical base for cooperation.

Like it or not, large entities working within one structure, whether it is government or corporate, inherently end up being competitive, even though they work for the same company or government. Getting these four cabinet ministries to work together toward a common end allows efficiencies of unusual scale.

Another advantage to industry is that the target group approach keeps the cleanup process from being distorted by competition. Because targets are applied to the entire industry, one company is never asked to do more (or less) than its competitors. And because each is equally responsible for the pollution created by the industry as a whole, they are more apt to share their cleanup technology.

By pulling together all aspects of a particular industry into one group that can work toward industry-wide solutions, the target group approach allows the Dutch to reach goals they could not otherwise hope to. A good example of how this works came about during the government's negotiations with the printing industry. One of the industry's environmental problems is that the ink it uses is toxic. A process-oriented solution would be to find a nonpolluting ink, but this is not something individual companies can achieve on their own, because they are dependent on the ink the chemical industry makes for them, what kind of printing machines exist, and so on. They would have to develop a new kind of infrastructure for a new printing process, and one company cannot do that on its own. However, by negotiating with the whole printing sector target group, including the chemical companies, the machine manufacturers, and the printers themselves, the government was able to encourage the industry as a whole to develop a solution, in the form of a nontoxic ink adopted industry-wide.

The target group approach is not just good for industry; it also has great advantages for government. For instance, regulators do not have to spend their time micromanaging, watching every move companies make and learning every detail of what they do. If a company can be convinced to do something voluntarily, the government automatically saves time and money, and the solution is probably much more efficient and effective than anything the government could come up with on its own.

The individual environmental plans each company is required to develop also take an integrated approach, which means that the companies will have to make an effort in all areas of environmental policy. This is particularly important for such issues as energy conservation, waste disposal, and soil sanitation, which are difficult for the government to control through a permit system. Most important, the long-term agreements will lead companies to opt more frequently for process-oriented solutions. Environmental policy will be more uniform throughout the country.

Another advantage of the target group approach for government is that it makes it easier to measure progress in each area. If the government had to negotiate with each individual company, it would not be able to tell if the reductions negotiated with each one were enough to solve the problem. The long-term goals for each target group provide a reference against which the plans of each company can be measured.

A Target Group Case Study

The primary metals industry agreement, the first to be reached under the NEPP, can serve as the example of the target group process and the types of goals it can achieve.[2] In negotiations over the agreement, the government was represented by staff from the ministry for the environment, the ministry of economic affairs, and the ministry of transport and public works. The provinces were also represented, as well as a union of the Netherlands' municipalities and other agencies.

In order to have a cohesive group that the government could work with, the primary metals industry formed a foundation that became the main negotiator for the businesses. The individual companies involved were all listed in the agreement.

The agreement is concerned with the role of this industry as a direct source of environmental pollution in the Netherlands, and the actions it will take to address this within the ten years after it goes into effect. In addition to provisions for reduction of emissions into air, water, and soil, it also incorporates policy with regard to energy conservation, waste materials, soil cleanup, external safety, smell, noise, and management systems.

In addition to the industry-wide plan, each individual company will develop its own green plan every four years. These plans will deal with the same issues as the main agreement, including energy conservation, the transportation of employees, and so on.

A discussion of the chemical cleanup entailed in this agreement would make a book in itself. But one example of the type of challenges these companies face in emission reductions is sulfur dioxide. In 1985, the industry released about 16,000 tons into the air. The goals for reduction of those emissions are a 35 percent cut by 1994, 75 to 80 percent by 2000, and 90 percent by 2010.[3] The amount by which each facility will have to reduce its own emissions is determined by the quantity of emissions it contributes to the problem.

There will be changes, breakthroughs, and difficulties during the implementation of all these agreements. Dealing with these uncertainties in a way that allows the process to continue is critical, so government and industry are planning ongoing discussions, consultations, and corrections. A consultation committee process has been established for that purpose.

One of the most important principles of the Netherlands' approach is trust. While the government will still have to enforce the laws and regulations on the laggards, it will not have to watch over every move a company makes. The phenomenal relationship between government and industry is illustrated by the fact that 65,000 businesses are voluntarily cooperating with the NEPP.

This kind of trust is working under the NEPP because industry in the Netherlands understands, and the public insists, that environmental decline must be reversed if society is to survive. Faced with that challenge, it is just common sense to cooperate with each other.

• • •

There are many ways to get companies moving in an environmentally sound direction, among them appealing to their enlightened self-interest and giving them financial incentives to adopt good practices and drop bad ones (some of these were discussed in chapter 8). Many businesses are heading that way on their own initiative. But green plans do what none of these methods can do by themselves: provide a comprehensive framework within which businesses and government can work together toward the common goal of environmental recovery.

In a recent speech to business executives in several states, J. J. de Graeff, a representative of industry in the Netherlands who has been extensively involved in implementing the business portion of the NEPP, noted five conditions that are essential to the success of any negotiation of environmental policy between government and industry. These are:

- maintaining a high level of political and social pressure for environmental improvement. This gives business the incentive to carry out the process.
- setting clear and achievable targets.
- recognition by both government and industry of the mutual benefits of working together.
- the continued threat of punitive measures by enforcement of existing laws.
- a reasonable level of trust between the parties.

Some of these conditions will be a challenge for many nations, including the United States, to meet. But they are far from impossible, and the success of the Netherlands can serve as a model for those who follow.

10

Building a Political and Social Base for Change

Because of the scale of change involved, the entire nation needs to work together if a green plan is to be successful. The green plans of New Zealand, the Netherlands, and Canada are partnerships among all members of a society to work toward sustainability, based on the belief that we are all collectively responsible for restoring and maintaining our environment. It begins with the individual: Each person must come to understand why it is important to make good environmental choices in his or her daily life. Stewardship of one's immediate environment must become a personal ethic.

This is true because the environmental problem is the cumulative effect of our individual actions as human beings. The consequences of millions of decisions made each day by people, companies, institutions, communities, and governments cause us to exceed environmental limits, drawing us like moths to the flame of future devastation. We have to change those decisions if we are to reverse environmental decline.

Fortunately, we have already been successful at creating environmental awareness. Most people around the world are already well aware of environmental problems, and most citizens of developed nations rank the recovery of environmental quality high on their list of goals. This framework of acceptance is a strong and necessary bulwark, providing support for the large-scale change that green plans will involve. If we were attempting to educate the citizenry from scratch, there would be very limited hope for success. As it is, the accumulated public knowledge about environmental issues has already served to motivate people to change.

The part that most nations have yet to accomplish is to get people truly involved in the effort, and not just in the traditional ways. Green plans require the individual citizen, as well as governments, to think in terms of processes and to look at the big picture. We have allowed ourselves to think that if we just recycle newspapers or sort our garbage we are demonstrating a commitment to stewardship, but stewardship within the green plan framework is a much broader vision of sustainability.

Finding a way to involve the millions of individuals who make up a nation is one of the ultimate challenges of modern times. It is the goal of almost every social movement and of a great many commercial efforts, from certain religious sects to political parties and membership organizations, on through to the latest attempts to sell soap. A whole industry – thousands of public relations firms and advertising agencies – has grown up around it. But while many attempt to change human behavior by capturing the public eye and mind, few accomplish it.

Traditionally, people who have wanted to promote progressive social changes in behavior have tried to do so by organizing on the grassroots level. This idea is cherished by all, as a kind of signal that democracy is working. But grassroots organizing has now come to mean preaching to the converted. Just as we cannot afford to deal with resources one issue at a time, we cannot expect to educate only the members of gardening clubs and make sustainability work: any appeal must be much more inclusive. What is really needed is the kind of effort that a political campaign involves.

To achieve the scale of change that is required, we must work with the organizations to which people belong. Broad acceptance of an idea is generally a result of getting a variety of social groups to support it. Churches, labor unions, professional associations, those interested in rapid transit or preservation – all need to be involved in a comprehensive solution to the environmental problem.

Such nongovernmental constituencies will have many roles to play in both establishing green plans and making them effective. Some will be roles in which everyone will take part, while others will depend on the work a person does, or the organization to which he or she belongs. From the perspective of the leaders and administrators implementing green plans, these roles deserve to be studied just as seriously as the technical or managerial actions that will need to be taken.

Before such nongovernmental constituencies can be built, however, the green plan process must be launched by the government sector that will guide it, the civil service sector, and be propelled to the forefront by a political leader or leaders.

Political Leadership

Visionary leadership from political officials can provide a great impetus for change. The public frequently needs a strong individual to serve as the focal point for an idea or cause before it will believe that change can really happen. Queen Beatrix of the Netherlands, Geoffrey Palmer of New Zealand, and Brian Mulroney of Canada provided such leadership in the countries that pioneered the movement for national environmental strategies. Because these countries were fortunate enough to have strong leadership from the beginning, they were able to build widespread public support for their plans, which is critical in the early stages of projects of this scale.

We have already seen (in chapter 4) how the Dutch were spurred to action by the famous 1988 Christmas Eve speech given by their queen, Beatrix. In New Zealand it was the commitment of former prime minister Geoffrey Palmer that propelled the country toward its massive reforms. When I interviewed Palmer in Wellington a few years ago, he told a story that explained why he was so committed to the idea of a comprehensive environmental strategy for New Zealand. Although his commitment was rooted in the love for nature he learned from his parents and teachers, it was his experience at the University of Chicago law school that really inspired him to take on a leadership role.

The University of Chicago is located on the south side of the city, in an area that has become poor and crime-ridden over the years. The university has maintained its academic excellence even as other parts of the neighborhood declined, but it is like a castle behind high walls, cut off from the world around it. Palmer felt very lonely and isolated there, afraid even to go for a walk off campus in the evenings because of the danger. His distance from the natural world was emphasized by the sight of steel mills in Gary, Indiana, not far to the south, belching pollution into the sky.

Faced with all this, he decided that if he was ever able to do anything for his own country, he would like to make sure that New Zealand never came to resemble the area surrounding the University of Chicago campus. When he became deputy prime minister he appointed himself minister of the environment and served in both capacities (later becoming prime minister) in order to get the green plan legislation on the national agenda.

As mentioned in chapter 6, Canada's prime minister, Brian Mulroney, was spurred to action in the midst of his 1988 campaign by opinion polls showing that the public – particularly young people – was very concerned about environmental quality. Mulroney promised the voters that if they elected him, he would start reversing the process of environmental decline in Canada. He won the campaign, appointed a new environmental minister, and the work of planning a national environmental strategy began. Public servants in the environment ministry and others went at it feverishly, delivering a plan that served as a starting point about nineteen months later.

It is possible that not every nation, state, or community will need this degree of inspirational leadership in order to get underway with a plan; the example set by these individuals may be enough to give other countries a clear path to follow. However, such leadership can be very helpful in speeding up the process of acceptance and in convincing the public to take part in the plan.

Strong political leadership can also be important after a green plan is underway; without it, the whole process can be derailed. In this case, it is not so much vision that is required as it is commitment to the idea and the courage to forge ahead. If at any point the government decides to back off and implement only part of a green plan, it will not work.

Building a Political Constituency

Political leadership alone will not be enough to ensure the passage and implementation of a green plan: a strong political constituency must also be built. Any powerful idea for change encounters opposition, as it should. In a democracy it is much easier to block an idea than to have it proceed, even if a majority of the public is in favor of it. That is a healthy condition, because not every idea deserves to be adopted immediately. A good idea will endure and win, but it should have to prove itself.

The green plan opposition includes those who are simply cautious about change, as well as those who believe there are no serious environmental problems that need changing. In order to free the idea to move ahead, some must be won over. The issue then becomes how to do it.

In my experience with interest groups, I have found that there are almost always people within the organization who disagree with the official stance on an issue. Even if the organization or sector remains opposed, you can still muster support from within its ranks. The trick is to convince those who agree with you to help.

It only takes a few voices of dissent within an organization to start people thinking, increase the scope of the debate, and in the end create internal controversy. If an organization has enough difference of opinion within its ranks regarding an issue, it may take a neutral position – often the most positive result one can hope for.

This is a phenomenon I first noticed within the labor movement, when as a government official I was trying to stop some unnecessary roads and dams from being built. I assumed that labor organizations would oppose this, because the roads and dams meant jobs for their members. But when I sat in on some meetings of the state labor groups, to my surprise I found that most of them were not in favor of the construction. The union whose members operated bulldozers and other heavy equipment was the primary source of the problem; many other unions, such as the Teamsters and those representing restaurant workers and carpenters, cared about the environment and opposed the new construction.

This is an important lesson to remember when building a broad-based political constituency: it is worth approaching even those you believe to be your opponents. Never assume that everyone in a group agrees with the leaders.

The political support of individuals and organizations will continue to be crucial even after a program has been passed into law. The people must be vigilant, persistent advocates of the government policies they believe in. No major program survives without the ongoing support of an outside constituency of citizens. Otherwise, next year a new issue rushes in and pushes the old one aside.

The importance of public support is crucial during the planning and implementation phases. This really is the essence of implementation: to create a political base of acceptance by those who live in and run the community, state, province, or nation involved.

Building a Consensus around Green Plan Goals
The consensus-building process New Zealand went through is a prime example of how best to create strong, permanent nongovernmental constituencies among both organizations and individuals. The country's drive for resource

law reform began when a former prime minister announced a massive development scheme that would have included damming some of the country's rivers. Part of the scheme included limiting public input into the process. The public rebellion over this created a political seedbed for overhauling the way the country dealt with development and the environment.

A new government led the reform effort. It was a huge task, requiring field hearings in every niche and corner of New Zealand, a toll-free number so any citizen could call in, television discussions – all over several years' time. But when it was through, they had reached a consensus regarding what could be done. This tremendous strength of having the people involved in the decision making is as important as anything else New Zealand has done.

New Zealand's example also demonstrates how important it is to build a strong public constituency for green plans, because it was public support that saved the plan from being scrapped early on, when the government changed before it had been passed into law. Changes in government almost always mean changes in policy: every newly appointed official, discreetly or otherwise, discards the ideas of his or her predecessor. Each new administration hopes that its own vision will become a permanent legacy.

So when the new party, which was more conservative than the previous one, took over in New Zealand, it suggested that the environmental plan be dropped. But the people were very much opposed to this: they absolutely wanted it carried out. Recognizing this, the new government established a committee to evaluate the plan, which made some changes and ended up improving it. Had the constituency not been there, the New Zealand plan would not exist today.

In contrast to New Zealand, Canada had to deliver its plan in one year, because of a promise the prime minister made during the election. That really was not enough time to get the people deeply involved in the whole process, to the point where they clearly understood and supported it. Although the Canadian plan is excellent, the time not spent in preparation will slow its progress in the early years, while the public goes through the learning process.

Building support for green plans and achieving consensus over their goals is no easy task. Businesses are always cautious about environmental policy, fearful of how much it will cost and how it will affect their ability to do business. Citizens worry about higher prices for products and personal inconveniences, such as having to take the bus instead of driving, or having to recycle.

Government planners and environmentalists both worry about protecting their particular piece of turf.

However, if each group can be convinced to think on a broader scale, a diverse, powerful coalition for change can be built. In the green plan nations, no one popular issue such as rainforests or toxics is allowed to dominate the bigger picture. With such a broad coalition involved, it becomes difficult for one segment of society to block any one portion of the plan. The plan in its entirety is stronger and more powerful politically than any of its components would be on their own.

Involving all levels of society in the process of creating a green plan does more than just build consensus over what actions to take. I have observed that, once a country has gone through the difficult process of sorting out what needs to be done to recover environmental quality, every institution and most individuals feel an ownership in the plan. Thereafter, the public is informed and committed to environmental quality as a principle for the nation.

Once they recognize that something is being done of a serious enough nature to actually solve the problem, the people realize that the gain will be greater than the sacrifice. They develop a commitment and a sense of purpose. Involvement on a personal level, such as recycling household waste, is tied into the larger, nationwide effort; everyone unites behind a common goal.

If the logging industry is faced with an environmental strategy that places the last 15 percent of the trees in parks and dictates that the industry farm trees rather than simply cut them, the loggers will be much more cooperative if they know that housing and transportation and the future for their children will be in much better shape. At the very least, they will not be as animated in their opposition.

Accepting Limits: The Citizen as Steward

Political support for a green plan and consensus over its goals, while quite important, are just the beginning: citizen involvement in the plan must go further. Achieving environmental recovery and a sustainable future will require a dramatic change in public attitudes. After several hundred years of believing that there were no limits to what we could do and consume, facing the fact that there *are* limits may prove to be our most difficult challenge.

Green plans provide the framework within which a society can examine its attitudes and assumptions. Working together, people can change their daily behavior by learning to make choices and decisions based on environmental considerations.

Here again, groups like labor unions, churches, youth and senior organizations, and professional associations can help to educate people. But the green plan countries have found that it also makes sense to appeal to the people individually, as citizens.

In the Netherlands, the government asked advertising professionals to help it with its public outreach. One of the things it did was spend several million dollars on a comprehensive environmental information campaign. The government realized that, although public concern for the environment was high in the 1980s, environmentally conscious behavior was lagging behind, because the environment was still too abstract a concept for many. It decided to adopt an integrated approach for disseminating information, based on a series of campaigns targeted to different sectors and issues. Various types of media are used, depending on who the campaign is designed to reach. All the campaigns are linked to provide a coherent whole by which the public becomes more aware that environmental problems and solutions are interrelated.

The broad general information campaign is targeted to the nation as a whole, and is designed to make environmentally friendly behavior second nature. Its purpose is to give people a sense of solidarity about working together to save the environment, showing how others are doing their part and demonstrating the positive effects of concerted action. The general campaign serves as a framework for other, more narrowly focused, campaigns, provides continuity between them, and keeps general awareness high.

The more focused thematic campaigns, also targeted to the general public, are aimed at raising knowledge and awareness of particular issues, such as climate change. Another campaign, aimed at consumers, focuses on certain types of products or actions. There is also a target-group campaign, aimed at behavior and practices in specific professions, such as construction.

The television ads, posters, and other materials that have been created for these campaigns are very effective; some have even won awards. One of the most popular posters presents three pictures of a forest, captioned "Yesterday, Today, Tomorrow?" "Yesterday" is a luxuriant, healthy forest; "Today" has trees whose leaves are dying; "Tomorrow?" is a photo of stark, dead trees with no leaves at all. The poster reads, "Acid Rain: Our Own Fault, Our Own Concern."

Reaching Out to the Public. This poster, entitled "Yesterday, Today, Tomorrow?" is part of the government's public information campaign in the Netherlands. It reads, "Acid Rain: Our Own Fault, Our Own Concern." Forests across Europe are dying due to acid rain. Photo courtesy of the Netherlands' Ministry of Housing, Physical Planning and the Environment.

One example of a very successful outreach program in the Netherlands is the recycling campaign. Citizens were asked to separate household waste into three different categories – organic, chemical, and "other." The program is working so well that the government is practically drowning in success, and has had to accelerate its efforts to catch up with the amount of organic waste being recycled. By 1994, every city in the country had a completely separated waste stream, as well as collection points on every corner for paper, metals, textiles, and glass.

Advertising and publicity campaigns can improve awareness, but that is only the beginning of the process. The real goal is to rise above our differences and create a shared purpose. People must become involved and committed to creating a tradition of environmental stewardship in their communities. This kind of tradition is not only possible, but already exists in various communities around the world. One of my favorite examples is that of Austrians and their love for the Vienna Woods.

Near the end of World War II, Allied forces came up against elite German troops when they attempted to take Vienna, and the ensuing battle was devastating. After the Allies had finally taken over the city, the commander issued an edict allowing the public to collect firewood for heating. But despite the fact that it was winter and deadly cold, not one stick of wood was taken from the Vienna Woods, the vast forest surrounding the city. That forest was so entwined with Viennese history, its cultural and artistic traditions, that the people would rather put on more sweaters and shiver through the winter than burn any part of their heritage. Their commitment to care for the forest had become a part of their culture. This is the kind of involvement we need to nurture in every community regarding its environment.

The story of the Vienna Woods shows that people can and do care deeply about nature and their environment. But that does not mean they will automatically do what needs to be done. In order to fully participate in a program, they have to believe that it is serious and will have a real effect. If we only deal with bits of the problem, the public knows full well we are tinkering.

Because green plans are so comprehensive and large in scale, people *will* believe in them, because they can see that they are truly serious efforts to solve the problem. The experience in the green plan countries has been a huge public outpouring of support. In New Zealand, the public involvement with this issue was greater than any other in the history of that nation. In the

Netherlands, the public voted overwhelmingly to support the NEPP after the prime minister had resigned because of the controversy surrounding it.

Some people stay on the sidelines because they are pessimistic about the possibility of environmental recovery. They feel helpless, believing that there is not much they can do that will have any effect. A part of the magic of these green plans is their potential to inspire. Because they are so comprehensive, they give society a unifying theme, a goal that everyone can get behind. For the Dutch, the idea of cleaning up for the sake of one's children and grand-children has been a powerful inspiration.

Environmentalists' Roles

There are several societal groups that have important, nongovernmental lead-ership roles; of these, business and environmental organizations are the two most essential. Without both, a green plan simply cannot happen. Because the idea of business cooperating on environmental issues is so new, and the meth-ods that green plan nations have developed for working with industry are so innovative, I have devoted a separate chapter entirely to that topic.

Nonprofit environmental groups will continue to play many of the same roles they do now. The most important is that of critic. Because of the nature of the political process, green plans will involve compromises. Trying to guide the whole nation, with its factories and schools and houses, its dumps and waterways, its transit problems – the whole, interwoven bundle – through a process of major change is an incredibly complex and difficult task. Envi-ronmental critics need to be on top of things, no matter how complex the is-sues are, to make sure that political compromises do not go too far.

The passage of a green plan does not mean that environmentalists should stand back and let the process take over. There is a definite place for their concerns; in fact, the whole program will benefit from their scrutiny. There will also be groups on the other side, pushing for less change and saying that national environmental plans are not necessary. While that group is trying to block the plan, the environmentalists will be on the other side, pushing for even stronger actions.

Critic can be a difficult role to play, because in order to be effective you have to be more informed than the experts you are challenging. But environ-mentalists have generally been out in front of the rest of the nation in under-standing environmental problems and possibilities. Interestingly, environmental leaders in the green plan nations tend to be better informed than their counter-parts in the United States. This may be because it was no challenge for U.S. environmentalists to stay ahead of recent governments that have been woe-fully ignorant about environmental matters.

Another important role for environmentalists is to get behind the idea and keep it going when political support is lagging. They really carry the responsi-bility of continuity for the program. Should a government attempt to back off from its commitment, the environmental leaders need be the voice stirring the public to demand more, not less, progress.

Finally, environmentalists will play a role in educating and involving the general public. Environmental groups that are independent of government must help people understand the dimensions of what is being gained, and in-spire them to stay involved. They must also make people aware of their own roles in environmental problems and solutions.

Some green plan nations have found that there is a fourth and somewhat new role for environmentalists: that of negotiator, sitting down with business, government, and other traditional interest groups to work out the programs. This may be the most difficult role of all, and not all environmental groups will want, or need, to undertake it. But in general, environmentalists will need to be more receptive to economic issues and the concerns of business. Eco-nomic policies have major environmental impacts that can be positive as well as negative: it is important to work together toward developing sound eco-nomic practices that incorporate good environmental policy.

We need to link the economy and the environment, to make economics green, because the world seems to still be working toward the traditional eco-nomic development goals. It will be a great achievement if environmentalists can turn that around and make sustainability the operating mode. That will mean working with past opponents in new ways, stressing negotiation over confrontation.

That does not mean that environmental activists should water down their defense of the environment, but a critic need not be an opponent. Environ-mentalists must work along with government, business, and other groups to shape the green plan and its programs. The involvement of at least a part of the environmental sector will make the plan better and stronger.

In the Netherlands, the government solicits input from environmental groups on all policy matters. Environmentalists sit on many advisory committees set up by the government, and at the Rio Summit, a representative from the environmental movement sat in on all the meetings of the Netherlands' delegation.

The Netherlands' branch of Friends of the Earth (FOE) is critical of parts of the NEPP, but supports it in general. FOE believes the program will improve as time goes on, and realizes that it is the environmental community's responsibility to help make sure that it does, by keeping the pressure on government and industry. FOE has published a number of fine papers critical of the NEPP, arguing that it does not go far enough.

The free voices of environmental leaders and their constituency are critical to a green plan's success. The environmental legacy of the former Soviet Union – a long list of tragedies that includes Chernobyl, the death of the Aral Sea, and the contamination of most of Eastern Europe – show how dangerous it is when a nation concentrates all its decision-making powers in one place, stifling the voices of advocacy and dissent.

•   •   •

There have been great moments in the history of nations, when a common cause united a country and propelled it forward. The hope emanating now from the green plan idea is one of those moments. But the magic in green plans goes beyond creating hope; in pulling a nation together, they create the potential for one of the highest human principles to emerge – that of working together for the greater good.

For whatever reason, society has chosen to write off many obligations that we owe to others and to the future. Our cultures are becoming very self-centered, and people are not happy with that. In the United States, we know that much of what makes our lives so comfortable has come from the commitment our predecessors made to the future, but we are not committing ourselves to the future in a like manner. Green plans give us an opportunity to change that.

There is a tremendous opportunity for such change in the United States right now. The vast majority of the voters, regardless of party, are concerned about the condition of our environment. However, for the past decade the country's leaders have not believed the environment was a priority. It remains to be seen what action the present leadership will take, but the public is ready, waiting, and hoping to be inspired and led in a direction similar to those that other countries are taking.

It is not often that whole societies come together to work toward a big, difficult goal, but when it happens, it creates an atmosphere of hope, a belief that we can truly make a difference. I was a child during World War II, but I remember clearly the feeling that absolutely everyone in the country had a shared purpose. That is the feeling one gets in the Netherlands or New Zealand now, with their programs.

11

A Greenprint for the United States

The time is ripe for the United States to adopt an environmental policy as new and sweeping as a green plan. It would be an historic proposal that would inspire the nation to solve the environmental problem. The rest of the world would certainly like to see it happen, because the negative impact of the United States on the world's environment will overwhelm the best efforts of the smaller nations, even if they are environmentally perfect. Just one example of the United States' impact is that it was responsible for about one-fifth of the world's total emissions of carbon dioxide from industrial processes in 1989.[1]

The idea of comprehensive, long-term environmental planning is not unknown in the United States, although it has not been tried here. Before Ronald Reagan's election, environmentalists and governmental officials alike had begun to think about the implications of regional and global environmental problems, and what might be done to stave off further deterioration.

In 1980, the Carter administration issued *The Global 2000 Report to the President*, an attempt to look at the big picture through a comprehensive assessment of global environmental and resource issues, projected over the following twenty years. *Global 2000* could have been the first step in a U.S. drive for sustainability, but after Carter's defeat it was ignored outside of the environmental movement.

Ironically, *Global 2000* had a significant impact on other countries. In conversations with environmental officials from New Zealand I have been told that it was probably the most important inspiration for that country's push toward a comprehensive policy. The writings of thoughtful U.S. activists such as David Brower and Barry Commoner – and, from an even earlier era, of

Aldo Leopold and Rachel Carson and others – have also had a greater impact on the rest of the world than they have apparently had on us. The thinkers in New Zealand and the Netherlands and other nations were reading these ideas years ago, and using some of them to forge ahead with progressive new policies. Many U.S. philosophers and activists played important roles in the movement that brought about the UN Commission on Environment and Development and the 1992 Rio Earth Summit.

So in philosophical terms, some of the green plan groundwork has been laid in the United States; at least some activists and government officials are aware of the issues and the important principles behind comprehensive thinking. And, as we saw in chapter 3, California made a preliminary step toward a green plan in the late 1970s with its "Investing for Prosperity" program.

But there has been little concrete action toward comprehensive environmental planning in the United States in the past decade. We have had two presidents (1981–1992) who believed the country's resources were still as unlimited as they were two hundred years ago, and that there would always be more forests to log, water to be used, gold and oil in abundance – that somehow we would be given a Garden of Eden in permanence, no matter what we did.

The days of no limits are largely over. We are moving, as have other countries in the centuries before us, from a nation that exploited resources without thought or worry for the future to one that must shift to sustainability if it is to maintain the quality of life for present and future generations.

The times are changing, but is the United States ready to change with them? If a green plan were proposed in this country, would it survive the political process? I believe the answer is yes, for a number of reasons. First, the public has long been aware of environmental problems and consistently supports strong environmental policies; almost 80 percent of the respondents in a 1991 Gallup poll said they consider themselves environmentalists. Young people are particularly aware of the serious nature of the challenge.

Concern for the environment is a very powerful, albeit latent, political idea, truly a gem waiting to be picked up by a political leader with vision. The pressure of environmental problems and the public's increasing interest in them has created a seedbed for change in the United States and other developed nations. The issues have grown far beyond the environmental movement as it has developed to date.

In addition, we can expect support on the part of government officials, who are employees of the taxpayers. Given a chance, these are professionals who care about doing their jobs well. But under the short-term governmental thinking of the Reagan-Bush years, they were stifled. That is changing; government officials at the federal, state, and local levels are already moving away from the thinking we have had in the recent past, and are beginning to prepare for the future. But the nation still awaits the political leader to pick up this idea and inspire it to act.

The Changing Business Environment

In addition to strong popular support and changing attitudes on the part of government officials, there is another reason a green plan is feasible for the United States – perhaps the most important reason of all. This is the changing attitude toward the environment within the business community. Over the last decade, our business leaders have also become more knowledgeable and aware about environmental issues. At the very least, most businesspeople and their public relations advisors realize that an anti-environmental stance is bad for their image.

A classic, early example of this occurred when it was discovered how damaging phosphates in laundry detergent were to the environment, and there was an instantaneous public reaction against the detergent manufacturers. Businesses moved quickly to remove phosphates from their products, and the smart ones have realized ever since that they should solve their obvious environmental problems before the public becomes involved and their image is tarnished.

There are economic gains for businesses and nations with progressive environmental plans, and there are penalties and problems for those that try to ignore the issue. Countries whose leaders have evolved better energy policies have gained a competitive advantage over the United States, which uses twice as much energy to manufacture the same goods as such trading competitors as Germany and Japan. This energy imbalance is untenable, both environmentally and economically. Becoming more efficient in our energy use, and the use of all our resources, will help us competitively, as well as bring environmental benefits.

All of this means that business is finding strong incentives to participate in comprehensive environmental planning. This type of planning not only makes them more competitive in the international community, but, as we have seen in chapter 9, simplifies the regulatory process and gives businesses long-term stability in the regulatory environment.

Other countries' green plans show us that business practices can be successfully adjusted to meet environmental concerns. The fact that a number of U.S. multinationals are moving ahead of the government suggests that they are applying the lessons they have learned abroad to their practices at home. The U.S. chemical industry, for example, has made major changes on its own. Dupont, Dow, and Monsanto are very progressive, and far different environmental managers from what they were a dozen years ago.

Another example is the electronics industry in California's Silicon Valley. Faced with increasing pressure from regulators and environmentalists, some electronics firms and other companies in Silicon Valley formed a cooperative association, called the Santa Clara County Manufacturing Group. Working together under its auspices, these companies have made dramatic changes. Between 1985 and 1990, they reduced their emissions of pollutants by 85 percent.[2] Their energy and water conservation programs have also been highly successful, as have their cooperative approaches to designing new CFC-free packaging made from recycled paper. Their example shows that U.S. businesses are capable of both undertaking comprehensive approaches and achieving dramatic results from them.

Of course, there are still challenges to face, one of the most serious being the difficulties of bringing environmental reform to the older, more polluting industries. Newer companies, like those in Silicon Valley, had advanced environmental technologies and practices available to them as they were developing and growing. They also tend to have younger, more forward-looking managers, who trained for their profession in an era of concern for environmental quality.

But the so-called heavy industries, such as coal and oil producers and the auto manufacturers, are mired in the old ways. This is partly due to the cost of refitting the old factories, but the attitude of their management, which tends to believe that the problems simply do not exist or are exaggerated, also contributes.

The case of United Technologies is a perfect example of the problems caused by such outdated thinking. United Technologies has a large facility in Silicon Valley, but its headquarters are in the East. For a long time, the headquarters paid no attention to the California branch's appeals for help in meeting the more severe environmental regulations the state had adopted. After years of doing what the home office told it to do, the California operation had been hit with a total of $12 million in fines for pollution violations.[3] At that point, the management back East finally got the message, and the California branch was allowed to make the changes it needed to comply with the laws. Now it has caught up and is very progressive.

Part of the problem, then, is simply transitional. I would estimate that one-third of our business leaders are environmentally aware, and more are changing daily. When the number passes 50 percent, as it should in the next few years, massive environmental reform will happen. There will no longer be a base of support for an anti-environmental business attitude.

The future hinges on those companies that have chosen the right direction, such as 3M, Dow, Dupont, and others. 3M, for instance, advocates a "beyond compliance" plan, which means the company sets goals for itself beyond the minimum the government requires. My favorite example is Patagonia, Inc., a sports clothing manufacturer that emphasizes environmental management throughout its operations. Patagonia products are designed for long-term use, and the company is eliminating most product packaging, among many other environmentally conscious actions.

Companies such as these understand what is happening in the world; that environmental concerns will be an integral part of our future. They have taken a leadership role, and are beginning to compare favorably with any of the corporations in countries where environmental quality is a high-priority issue. The next step is to convince these companies to actively oppose the outdated attitudes of some in the U.S. business community, and to be a stronger political voice for progressive environmental policies.

Building Momentum from the Ground Up

As we have seen, there is a large constituency in the United States, made up of citizens, government officials, and business leaders, that is ready to get behind major environmental policy changes. The 1992 Rio Earth Summit seems to have sped up the process, inspiring people and governments around the world to respond to the challenge of creating sustainable societies. For example, in May of 1993 the governor of Kentucky held a sustainable development conference called "From Rio to the Capitols," to which all but one of the states sent participants. The topic was comprehensive strategies for sustainability on the state level.

In addition, the Clinton administration established a commission to study sustainable development and gave it two years to define precisely what the term means to the United States. The commission will also formulate a national strategy in response to the Rio Summit's Agenda 21. Predictably, the nation will have gotten far ahead of the commission before that comes to pass, but small steps are at least being taken.

Small, because until major political leadership is behind an idea, government is not in a position to manage such a large and complex concept. It does not appear that we will have that sort of leadership at the national level for some time, but a great deal can be done at the state level, or at city levels, for that matter. Some communities are already talking about creating their own green plans.

In the last few years some of the states have started to move in the direction of more comprehensive resource planning. Idaho's Mid-Snake River Study Group, established in 1989, is composed of government and public representatives from the counties through which this portion of the Snake flows. It was formed solely to improve water quality in the Snake, which is as important to its region as the Mississippi is to its watershed. However, its emphasis on comprehensive ecosystem management through cooperation between all stakeholders could provide a useful model for regional, watershed-based green planning. Its Coordinated Water Resource Management Plan, which recommends legal and regulatory changes and would help coordinate the flow of local, state, and federal resources, was at the time of this writing being considered by county commissioners.[4]

A number of states are far enough along that it is quite possible they could adopt a green plan within the next few years; New Jersey, Washington State, and California are potential candidates. But activity in most states has been limited for a variety of reasons. The most important is that the problem has not been given priority by our current political leaders. If they did consider it a priority, they would translate their concern into an action plan and budgeting. That leads to the second problem, which is a lack of funds. There is always a lack of funds, because all the money is always spent. But if priorities are changed, then money can be shifted from the old priorities to the new. Another problem for the states has been changeovers in state governments that have cut short the programs initiated by previous administrations.

There is nonetheless something to cheer about in terms of emerging environmental programs in the United States. Attempts are being made in the direction of truly comprehensive environmental planning in several states. I believe these initial state efforts will catch the attention of national leaders, and the movement will begin to roll.

The EPA is well aware of what is going on in the rest of the world, but it cannot make fundamental changes in environmental policy on its own. It has attempted to begin the process of comprehensive planning by funding comparative risk assessments in the states, which are intended to be preliminary steps in a strategic management process that will lead to long-term, comprehensive resource management plans. A number of states are working on or have completed their assessments, and several are ready to move beyond this stage.

The major advantage to the risk assessment process is that it can be a means of pulling together various groups in each state to think about and discuss the key issues affecting them. The best of the state groups have functioned in much the same way as the Canadian Round Tables, bringing together government officials, scientists, and other so-called "stakeholder" groups – business representatives, environmental groups, labor unions, community activists, individual citizens, and others – and sitting them down at one table.

Washington state has adopted a master plan for its resources, called "Environment 2010," which grew out of a risk assessment. In 1991 the state enacted major environmental legislation that came directly from the 2010 process, including air quality laws, water and energy conservation strategies, environmental education programs, and oil spill prevention policies. One of the major policy changes Washington enacted because of Environment 2010 was to pass managed growth laws. The 2010 process recognizes that the environment is a system, and covers cross-media issues as well as traditional concerns.

One of the most environmentally pressured states in the union, New Jersey, is also one of the most environmentally progressive. New Jersey has some of the country's strictest environmental standards. In addition, the state's planning commission has developed a comprehensive environmental and land-use strategy with both broad and specific goals covering a spectrum of concerns. The State Planning Act that called for the plan's development also established an ongoing monitoring and evaluation program, in order to adapt the plan to changing conditions. This is the only instance in the United States so far of continuous monitoring and reevaluation of environmental policy.

Vermont has also passed legislation that has set it on the road to more comprehensive environmental planning. This legislation created a state environmental board and district environmental commissions with the power to establish comprehensive development and land-use plans. The board is responsible for issuing land-use permits, and cannot issue a permit for development unless that development meets all the environmental criteria outlined in the act. Vermont is also in the process of completing a comparative risk assessment.

Florida's Environmental Strategic Plan (ESP) will, when fully implemented, provide for long-term strategic planning, program planning, budget development, and environmental and program evaluation. The ESP provides the overall direction for the next five to ten years, strategic implementation plans that detail the objectives and activities for the current year, and comprehensive evaluations of environmental performance. The state is currently in the process of doing a risk assessment.

Oregon initiated a risk assessment-based strategic planning effort in 1989 to come up with a comprehensive plan for the state's future. Called "Benchmarks," the plan set five-, ten-, fifteen-, and twenty-year goals in three categories: people, economy, and quality of life. The quality of life category includes approximately thirty-five benchmarks dealing with environmental issues, among them carbon dioxide emissions, wetlands and forest preservation, erosion control, transportation, and water quality.

Since 1989, Minnesota has had a trust fund for investing in its resource base. In 1991, it began a process called "Minnesota Milestones," based on Oregon's "Benchmarks" program and intended to create a sustainable development plan for the state. With citizen input, this program has identified priority issues for the state and adopted a vision statement for the future based on these issues. It has set specific numerical goals for most of the issue areas – for example, that average annual energy use per person will be reduced by 22 percent by the year 2020.

Connecticut has an environmental plan, Environment 2000, which is revised every five years. Each new plan identifies critical environmental issues, sets goals in each issue area, and identifies strategies for reaching the goals. The plan serves as a guideline for future legislation and budgeting procedures.

A few other states are in various stages of developing, planning, and implementing similar projects. Pennsylvania and Louisiana have done risk assessment-based projects and developed environmental master plans, but changes in administrations have stalled further progress.

The preliminary groundwork has at least been laid in these states; they are familiar with and receptive to some of the concepts that are important to green plans. What they need now is information about the existing green plans of the Netherlands, New Zealand, and Canada. These successful plans can provide practical, hands-on answers to the questions the state planners will have as they tackle the business of creating a green plan.

With these developments in the states, as well as some others, I believe that this country will move forward. Those states that have started will pull the rest along with them. But green plans do not happen overnight: as the experiences of the green plan nations have shown, a solid foundation must first be built if change on such a massive scale is to be successful.

A Federal Framework and Funding for a United States Green Plan

What are the key hurdles, the early challenges that will have to be met in order to increase the scale of green planning in the United States? Funding is one, but it is important to remember that good management brings large savings in its wake. Simply improving our system of regulations, subsidies, and incentives will greatly increase management efficiency, and that in itself can free up funding. The polluter pays principle, major energy and water conservation programs, and recycling and solid waste programs all save money.

Most green plan activity will take place at the state level, but the states will need both financial help and guidance from the federal government. Therefore, we will need some sort of framework at the federal level that provides this help but allows the states to make their own decisions. The federal government has often successfully involved itself on the local level through programs such as Community Block Grants, which currently channels federal funds to the states for such programs as Women, Infants, Children; Head Start; and the program that helps low-income households pay for home heating. Although the federal government provides financing and manages these programs, decisions are made at the state and local levels.

Such a program could easily be set up for state green plans. Its implementation might look something like this:

1. A federal commission on green plans is formed. Appointed members represent the House and Senate, president's cabinet, and Department of Justice, among others. This commission distributes funds to the states for green plan development.
2. Each state creates a green plan commission that receives the seed money, defines the state's environmental problems, and outlines the priorities.

3. The state commission prepares a proposal to pay for green plan development, based on the state's short-term needs and development of long-term strategies, and submits it to the federal commission.
4. The federal commission allocates substantial funding each year to states on this application basis.
5. States would transfer funds to communities or regions to prepare local green plans, based on each state's priorities and guidelines.

The question will still be raised as to where all the money will come from; this was also the case in the countries that have already adopted green plans. A great deal of money can be made available for these programs through, for example, a gas tax, which would also give the public an incentive to drive less. The United States currently has some of the lowest gas taxes in the industrialized world; gas prices here do not come close to reflecting the environmental costs of driving.

A portion of the funds generated by a gas tax could be used to ease the burden it would create for the poor, and another portion to provide conservation incentives to the oil industry. Such incentives have worked well with, for example, the energy utilities; California's experience shows that they provide many environmental and economic benefits, as well as securing the active cooperation of the industry involved.

Overcoming Institutional and Psychological Barriers

The greatest challenges we face in establishing a U.S. green plan, however, are neither political nor financial, but institutional and psychological. Green plans require a shift in the way we think and act, individually and in business and government. Continuing, high-quality education will be one of the most important factors in achieving sustainability. To understand the more complex management approaches it takes to run a nation sustainably requires an informed and educated populace.

The public needs to be able to understand, for example, the difference between a project and a process – a distinction that can be quite difficult to grasp. The complexities of comprehensive environmental planning require a great deal of time and thought, and the development of skills related to long-term, process thinking. By contrast, we tend to prefer quick, simple debates that provide short-term answers, and are used to thinking of projects, not process. Political campaigns with thirty-second sound bites from the candidates are perhaps the worst example. A high level of education is the only way to reverse this situation. The literacy rates of the Netherlands and New Zealand are among the world's highest, close to 100 percent; while the U.S. literacy rate is high, it does not approach these countries'.[5] We cannot afford to let this situation continue.

It will also be a challenge to train good resource managers. To learn a subject well, we often focus upon it to the exclusion of other subjects. But the increasing specialization in science means that university departments, as they currently operate, are not able to provide the diverse training a student in this field requires. These departments are, by and large, still teaching in the same way they were when I was in graduate school studying natural resource management. The university I attended had two departments concerned with water: watershed management and water quality management. There was very little cooperation between the two, to the point where students in one department often were not able to take classes in the other.

Modern resource management training will have to move away from these reductionist traditions of academic science and toward a broader acceptance of the interrelationships that exist in nature and society. The difficulty of thinking beyond one subject to the relationships among a multitude of subjects is the greatest challenge American learning institutions face in preparing green plan managers. Opening our minds to a broader way of thinking is going to require a new approach.

Because many of the management techniques important to comprehensive resource planning have been adapted from the business world (see chapter 8), there needs to be some synthesis between the schools of business management and resource management in the future. The emphasis, however, should be on management rather than science; what we need is managers who know how to use scientific information.

Reviving Planning in Government

Even more important, however, may be the need to educate our political and institutional leadership. One reason that the United States has made so little progress toward comprehensive resource management is that long-term planning has fallen out of favor when it comes to dealing with our public affairs. It is interesting that businesses rely so heavily on it – business plans are standard practice – and yet we tend to believe that it will not work for government.

Again, Americans have never believed there were limits to our resources or space. Why should a country devote time and resources to planning if there are no limits? The Dutch ran out of their timber centuries ago, and have had to plan around it; they became traders because of their shortage of resources. They had to reclaim much of their land from the sea, and their tiny country has been besieged for years by the pressures of population density. As a result, they have long had to plan in order to manage their problems. We in the United States are finally beginning to realize that we are facing the same reality, and that what we do now will have great consequences for the world our grandchildren will live in.

At times planning has been popular in the United States, most recently in the 1960s. Since then, it has fallen out of favor for two reasons. The first was a succession of national administrations hostile to the idea; they did not believe that planning for long-term use and productivity was necessary. But since business itself has always emphasized planning in its own operations, the idea has never entirely died out here.

The second reason is that "planning" has increasingly been used as an excuse for not making a political decision. Most plans have been made to be put on shelves, not to solve problems; politicians hope that if the planning process goes on long enough the people will forget about the issue at hand. Planners have erred in being a tool for that kind of inaction.

Planning documents have also often been severely compromised by trying to be too many things for too many people. Most communities' master plans from the last twenty years are so full of exemptions and loopholes that they look like Swiss cheese. The result is that the public lost faith in the process, and in essence, government planning has died out.

But now that the accumulation of our problems is reaching a life-threatening scale, we may turn to it once again, as other industrialized nations are doing. If planning makes sense for business, it also makes sense for government, to ensure that tax dollars are properly spent and services provided in the most efficient way. The successes of the green plan countries bear witness to the fact that the failures of planning are not due to any inherent flaw in the concept, but rather to defects in the way it is carried out.

Developing even a few of the key green planning measures – a set of national environmental priorities and goals and a set of indicators to measure progress, along with the requirement of regular public reporting – would be a great improvement over our current situation.

The Problem of Scientific Integrity

Finally, one of the most crucial – and difficult – challenges we face in the United States is changing our attitudes toward and our use of science in policy making. There is a pall over scientific integrity in this country, largely because our adversarial legal system has warped the idea of scientific evidence. This has led to a phenomenon some have dubbed "dial-a-scientist," referring to the fact that you can hire an "expert" to say almost anything and refute almost anything. The integrity of science in the courtroom, and in decision making in general, has declined precipitously.

When I was California's secretary for resources, I came head to head with this phenomenon when I called my first hearing, on a toxic agricultural chemical. I called the hearing because the scientists in the resources agency brought in research from universities and elsewhere confirming our suspicions that this substance was detrimental to health when applied to food products.

The chemical companies that produced the chemical brought their own scientists to the hearing, who began pecking away at our data, focusing on minutiae and quibbling over irrelevant details. In the end, even though we were certain of our data, they raised enough doubt that the legislative committee would not make the recommended changes.

This is standard practice in the United States. Nearly every decision we made in the resources agency resulted in a lawsuit, based on "scientific" challenges to our information. This is equally true in Washington, D.C.; former Interior Secretary Cecil Andrus has told me that every decision he ever made ended in a lawsuit being brought by one side or the other. In 1993, the Supreme Court even ruled on how "expert" a scientific witness had to be in order to testify in court.

We need an applied science agency that is independent of government, and which has great integrity in the field. This was, of course, what the National Academy of Sciences (NAS) was intended to be. But the Academy damaged its own reputation when it became politicized during the Reagan administration, a time marked by constant attacks on the environment by the president and his appointees.

On one occasion, Reagan went on television to challenge scientists from the EPA, which was attempting to impose regulatory controls on a particular toxic substance. After criticizing the EPA's scientists as inept bureaucrats, the president said he was calling in the NAS to do further research, for which the government would provide funding. That act on the president's part did serious damage to the NAS's credibility as an independent, nonpartisan source of scientific information.

It is troublesome that many of our environmental problems are compounded by the lack of credible scientific information. How do we correct this? First, we need to change our attitudes toward the role of science in society. Scientists can give us information about a problem or a solution, but they cannot tell us how to act upon that information, because policy decisions inevitably require us to consider a range of issues outside the realm of pure science. To ask scientists to make such decisions is wrong. The issue is to divorce science from the political process. Their proper contribution to the debate is to provide the best and most complete scientific information possible to policy makers and to the public in general, with whom the final decision should rest.

We should establish an independent scientific institute with unquestioned integrity and strong credentials in the applied sciences, similar to the Netherlands' National Institute of Public Health and Environmental Protection (RIVM). Although the Netherlandic government does fund RIVM, the institute is independent, objective, and highly respected. It was this institute that researched and wrote *Concern for Tomorrow*, the report that galvanized the Dutch to action and set the environmental goals for the country. RIVM also conducts regular reviews of the NEPP.

Within RIVM, the goal is scientific cooperation and consensus over methods and interpretations. If consensus cannot be achieved – which occasionally happens, especially when two models based on opposing theories are raised – then both alternatives are presented to the government. RIVM has nothing to do with setting policies, following the Netherlandic practice of making a sharp distinction between science and policy. This keeps the institute from being politicized, and preserves its scientific integrity.

RIVM does its work so well, with such honesty and openness, that its conclusions are generally accepted and it is able to withstand political pressure regarding its findings. None of the information RIVM works with is confidential, and its methodologies are available for study and critique. That means that if it is asked by Parliament to do a study, and a government agency disagrees with the conclusions, RIVM gives the agency its data and asks it to have its own analysis done. The general public also has access to RIVM information. This does away with a great deal of uncertainty and debate.

The independent work that RIVM does is essential to much of the smooth operation of the government and its relationships with industry and the citizenry in general. The Dutch have been able to move so far ahead so quickly in their environmental policy largely because they can agree on the targets and goals set by RIVM.

• • •

Important ideas for change have often begun in smaller, more culturally homogenous nations with a democratic tradition, most likely because it is easier for smaller nations to agree upon dramatic change and to work out the difficulties of implementing it. Once they have done so, other nations can more readily follow their example. I believe this will be the case with green plans: now that countries such as the Netherlands, New Zealand, and Canada have blazed the trail, the United States, and the rest of the world, will find it easier to follow suit.

It *must* be the case – the world has no choice, if it is to survive. The green plan countries know that their efforts will not be enough if the world does not join them. The United States, with the world's largest economy, will send a powerful message to other nations by making a virtue of this necessity and forging ahead. In the 1960s and 1970s, this country was a world leader on environmental issues; now we have the opportunity to be so again.

Green planning is an idea waiting to be discovered by political leaders. Every candidate for public office dreams that he or she will make a difference in government and society, and looks for new ideas that will help bring this about. The difficulty lies in translating ideas to reality, and most politicians are reluctant to promote an untested idea, particularly when it means a significant change in the status quo. Eventually, green planning will be taken up by leaders around the world, and it will be immensely successful. It *has* been tested, and has shown its worth. The idea is gaining momentum, and as it does, it will find the leaders to champion it.

Afterword

When I first began writing this book, I knew that developments in the world of green planning would outpace its publication. In fact, since that time many nations, regions, and cities have begun the process of altering their environmental policies to reflect green plan principles. Many do not use the name "green plan" to describe what they are doing, but the name is not important as long as the basic principles remain the same.

The environmental gains made by nations that have adopted green plan principles are remarkable, but one of the most striking benefits has been an unexpected improvement in governance across the board. This is an impressive accomplishment, considering that improving the governance of a nation is among the rarest achievements of our time. Green plans require negotiation and cooperation among many sectors of society, and it is this process of cooperation that has yielded the unexpected benefits in governance. For example, the Dutch have found that the process of negotiating and implementing their green plan has led to cooperative efforts in other policy areas; the overall result has been more efficient and effective government.

The main challenge to the green plan idea is that it is still so difficult for many to grasp the big-picture, process approach. We are trained to think simply, in linear ways, while natural systems require us to use systems thinking. It is easier for us to take things apart and look at the pieces when what we need is to comprehend the whole. Likewise, there is a tendency to respond to green planning by enthusiastically adopting parts of it, but while the parts are undoubtedly good and worthwhile, it is the whole that is important.

For example, the White House and the EPA have pulled out one component of the Netherlands' green plan and used it to develop a number of programs, such as Project xL, the Common Sense Initiative, and the Environmental Leadership Program, which encourage companies to move "beyond compliance" with existing regulations. This is a positive step, but misses the fundamental point of green plans, which is that only a large-scale, inclusive effort can provide the broad base of public support required if society is to solve the problem of environmental decline.

Despite such challenges, a great deal of progress has been made around the world. The green plan countries have largely stayed on target, and many of the nations that were on the path to green plans have moved even closer. One of the most impressive of these is the European Union's comprehensive, integrated pollution prevention and control program. An innovation of this plan is in the permitting process for new industrial sites in European Union nations: permits are now granted on an integrated basis and must consider the effects on air, land, and water, and on neighboring nations. The program also requires member nations to share information and technological advances.

A number of other nations are now closer to joining the green plan ranks. This has been particularly true of developing countries, from Mexico, Colombia, and Peru to Bulgaria and Latvia. However, I am pleased to report that there has also been a great deal of progress in the United States. In fact, one of the programs I am most excited about is Minnesota's; as of this writing, Minnesota and Mexico have emerged as two leaders among the governments developing green plans.

The following are some of the important recent developments in the world of green planning.

The Netherlands and New Zealand: Green Plans in Action

The Netherlands' green plan effort—the planning and negotiating, the legislation and its implementation—is truly a historic accomplishment. Now seven years into their twenty-five-year plan, the Dutch have met most of their goals and objectives surprisingly well. Perhaps most important are the changes that green planning has brought to the entire operation of government in the Netherlands. One is that the concept of managing for total environmental recovery has been established as the operating principle within the national and local governments, and government officials are now comfortable with thinking and managing comprehensively. Another is that the process of cooperation on environmental management that was developed between government and industry has led to improved governance in other policy areas as well.

182

The Dutch continue to develop innovative programs to help them achieve environmental goals. For example, to encourage the development of alternative energy sources such as wind, the government has granted tax-free status to earnings on investments in "green" energy production. In addition, the government recently imposed a new carbon tax on fuel in the hope that increasing the costs of driving would encourage people to use mass transit, thus reducing carbon dioxide emissions. Some of the money raised by the tax will be used to improve public transit and to create other incentives to reduce carbon dioxide, such as loans to businesses for replacing old, fuel-inefficient equipment.

The support of the business community has been key to the success of the Netherlands' green plan and will continue to be in the future. There are still opponents to the green plan within the business community, and they can be politically powerful despite being in the minority. Overall, however, business support for the green plan has been very strong.

One unexpected benefit to the business community has been a change in public attitude. Ten years ago, opinion polls determined that the public's reaction to the business sector was negative; most citizens did not trust it. Now the polls show that the public has more trust and respect for those companies that have cooperated—and sometimes even led the way—in the green plan effort.

New Zealand's Resource Management Act (RMA) is also making great progress, although its sweeping changes in the law and even in the structures of government are still in the process of being implemented. Staff members from Resource Renewal Institute visited New Zealand in 1996 and came away satisfied that the majority of the population—and of the country's institutions and industries—is proud of the RMA and committed to its success.

Such large-scale changes will always involve some difficulties, however, and implementation of the RMA has not been entirely smooth. One example is the shift to one-stop permitting, as opposed to the old multi-agency, multiple-permit approach: although it is moving ahead, initial expectations were overly simplistic. The funding devoted to the establishment and administration of the new system was not adequate to the task.

As in the Netherlands, New Zealand's green plan has faced opposition from a number of industrial sectors. These are in the minority and tend to be those that have little long-term perspective, such as the mining, fossil fuels, and real estate development industries. New Zealand's mining sector managed to secure a court decision that makes environmental groups responsible for court costs in lawsuits brought against the industry. This has had a chilling effect on lawsuits.

There is no question of New Zealand backing away from the RMA, but it would help speed the process if the government would create a vehicle to support environmental groups and other nonprofits. Most industrialized nations allow tax deductions for charitable contributions to nonprofits; New Zealand would greatly benefit by adopting this policy. If government and industry are two legs of the stool, nonprofits are the important third leg, providing the criticism that helps keep government policies honest and fair.

The other major green plan covered in this book, Canada's, is still in something of a holding pattern, due to the change in governments that was taking place as the book was written. The administration of Prime Minister Jean Chretien, who succeeded Brian Mulroney, has turned its back on the green plan in the form initiated by his predecessor. However, it appears that the country will follow some sort of sustainable management approach. The government recently created a Commissioner of Environment and Sustainable Development post, and in the same act required that all ministers present sustainable development strategies for their departments to Parliament. In addition, several of the Canadian provinces have moved ahead with their own programs. British Columbia, for instance, has put together a green plan approach that has surpassed the original objectives of the national plan.

These changes brought about by the Canadian political process may serve as a test for the ability of green plans to adapt and survive over the long term. Although the original framework has been dropped, many of the green plan's key components have been maintained, and the desire for sustainability still ranks high with the Canadian public.

Progress in the States

Since this book's publication, several states have made significant progress toward adopting a green plan; foremost among these are Minnesota and New Jersey. Minnesota emerged as the leader in 1996, when it passed legislation that defines sustainability for the state, directs state agencies to integrate sustainability principles into their operations, and further directs the state Environmental Quality Board to oversee the process. The legislation requests that the board develop principles, to be ratified by the Minnesota Roundtable, for state agencies to use in their program areas. In addition to defining sustainability, the law directs the Minnesota Office of Strategic and Long-Range Planning to create and publish a guide to local sustainable development planning that includes model ordinances. One goal of that directive is to broaden the perspective of local planning agencies beyond the narrow technical approach they traditionally employ.

Another law Minnesota passed in 1996 allows businesses that cooperate in sustainability programs more flexibility in how they achieve the environmental quality goals set by the government. The effect of this law will be to shift the state away from the old regulatory practice of micromanaging business activities.

These new laws grew out of a process that began in 1993, when the governor and the Environmental Quality Board appointed 105 citizens to participate in a sustainability initiative. These individuals were divided into teams covering seven issue areas: settlement, manufacturing, agriculture, energy, forestry, minerals, and recreation. After three years, each team made its report; the reports were summarized in a booklet entitled *Challenges for Sustainable Minnesota.* The public review draft of this document is available on RRI's web site (see p. 4).

The process of passing the legislation was just as important as the legislation itself; the debate surrounding the passage of the laws led to the education of officials, the media, and the public. With these laws, Minnesota has become the first state in the nation to define sustainability in working terms. It has courageously grasped the green plan vision and has begun to implement it. Now the rest of the country has a model that has been discussed and debated in the heat of the political arena.

New Jersey began to express interest in the idea a few years ago, after the election of Governor Christine Todd Whitman. Members of her administration visited the Netherlands in a group that also included New Jersey foundation representatives and other interested individuals. After observing the Netherlands' green plan in action, the tour participants enthusiastically proposed a similar program for New Jersey. Since then the state has integrated a number of green plan principles into its environmental management policies, and has established a task force to determine how the Netherlands model can best be applied.

A number of U.S. cities have also become interested in sustainability policies, and a few have even begun to implement them. San Jose, California, and Chattanooga, Tennessee, are two of the most active. Not long ago, Chattanooga was believed to be one of the most polluted places in the country.

In the last few years it has completely reversed direction and in 1996 received an award from the President's Council on Sustainable Development (PCSD). These awards were given in recognition of exemplary, pioneering efforts in sustainability in the United States.

San Jose, which has suffered from uncontrolled development in the past, has taken control of the development process by establishing a boundary around the city beyond which development is not allowed. The San Jose city government has also finished preparing a series of excellent sectoral plans on such issues as water and transportation policies. It is now in the process of creating a green plan–type environmental and development policy structure that will help it fill in the gaps in other issue areas and link them all in a comprehensive, integrated, systemic way. With these steps, San Jose has taken the lead among U.S. cities in terms of potential for achieving a green plan.

On the national level, government officials and business and community leaders have been showing increased interest in green plan principles and ideas. One indication of the growing momentum was the creation of the above-mentioned President's Council on Sustainable Development, the U.S. response to the UN Earth Summit in Rio. The PCSD's report to the president, issued in 1996, was a much-needed step toward green planning on the national level. It is very good for a compromise document and some parts of it are particularly strong, such as the section on population, which marks the first time the federal government has formulated a policy regarding this issue.

The report also proposes to do away with the crippling subsidies that are degrading our public land resources and with the poor public land management practices that were established in the last century. On the other hand, its lack of a section on energy is a notable defect; this may have been due in part to the opposition of two oil industry executives who served on the council. The president's initial response to the report was to agree to implement the portions that could be accomplished by the executive branch of government.

As part of its report, the PCSD presented fifteen awards to organizations and companies that represent the best examples of sustainable management in the United States today. We at the Resource Renewal Institute and the State of California's Resources Agency received one of these awards for the Investing for Prosperity program in California (described in chapter 3 of this book) and for our green plan work. It was a great honor for us to be chosen out of an initial field of 1,800 nominees; but more important, it is an indication that green plans are beginning to gain the institutional support they need in the United States.

I believe that in the future, green plans will become a powerful political factor in the United States because of their appeal to a broad coalition of interests. With their focus on quality-of-life issues and efficient management, green plans have an appeal beyond the traditional environmental realm. For example, "wellness" advocates and others in health care are interested in green plans because they address environmental problems that cause disease and illness, such as air pollution and contaminated drinking water. To solve such problems both improves the nation's health and cuts health care spending.

Mexico: Facing the Challenges

We were excited to learn late in 1995 that Mexico was completing a national environmental recovery plan. Entitled the National Environmental Program 1995–2000, this green plan is a remarkable accomplishment that puts Mexico well ahead of many other nations in terms of environmental policy. It is an impressive show of leadership for the Americas. We were particularly pleased to learn, in conversations with officials from the environment ministry, that RRI's green plan conference publications and material from our Internet library served as sources of information for their preparations.

A comprehensive environmental program became a priority for Mexico in part because of its long involvement in international environmental affairs, and also because of the more recent demands of international trade. The country's participation in the international arena has been thoroughly discussed in the Mexican legislature and the media, raising public awareness of the issues and familiarizing both the public and government officials with international developments in environmental quality management.

There were a number of steps involved in the formulation of the Mexican green plan. First there were broad grassroots consultations for the National Development Plan, of which the environmental program is a component. The National Development Plan was presented to the Congress six months after President Ernesto Zedillo took office.

Then came a series of meetings between governmental agencies to discuss general objectives, methodologies, and coordinating mechanisms for the environmental program. Next, the government held consultations at the state and national levels with private and social sector organizations to discuss the proposed plan and its objectives. These organizations included chambers of commerce, labor unions, academic and professional associations, environmental groups, and peasant organizations.

The government also sought ideas from its Sustainable Development Council. After integrating the council's comments into its draft of the plan, the environmental ministry took steps to make the plan fully compatible with other national programs and policies, from poverty and education to fisheries, energy, and water.

At this point, the secretary for the environment made an interesting political move, insisting that the draft environmental plan be opened to a second round of public consultations and hearings. This took another year, but strengthened public support for the green plan.

From what I have seen of Mexico's plan so far, it is a major step toward the sort of comprehensive environmental policies that the Netherlands and New Zealand have developed. Mexico has not yet committed the same level of resources to its plan as have those countries, but it has tremendous potential. A great deal of progress was made just by going through the difficult political process that had to take place in order for the plan to be launched. A developing nation in the midst of difficult economic times, Mexico will meet numerous challenges as it struggles to implement its green plan; many other nations will be watching and waiting to judge the results.

Singapore: A Sparkling Example

When I wrote the section of chapter 7 that pertains to Singapore I had not been to that country in several years, and so had not had the opportunity to observe in person the amazing transformation it has brought about in terms of quality of life for its citizens. On a visit in January 1995, I saw a city-state that had been transformed into an impressive, environmentally sparkling, model green plan nation. Singapore's impressive economic advances have been matched by its commitment to environmental sustainability, public health, and overall quality of life. It provides an excellent example for other major cities around the world, many of which share the same problems Singapore once faced—population density, air and water pollution, problems of waste disposal, public health issues, and the need for efficient transit, among others. Like Singapore, these cities will find that a green plan can help them revitalize a decaying environmental base.

The success of Singapore's green plan is due in part to the fact that the country's former, long-time prime minister, Lee Kuan Yew, was committed to the environment from the early days of his administration. His leadership in this area brought his country from an impoverished, environmentally stressed nation to an industrialized power whose citizens enjoy a very high standard of living, including high environmental quality.

Singapore learned early on that environmental concerns are integral to any concept of quality of life, and so managed to avoid many of the mistakes made by nations that industrialized earlier. The green plan and its action programs have helped Singapore achieve even greater results in its drive toward sustainability. In recent years, the country has ranked as one of the top three industrialized nations in the world in efficiency of power generation. Singapore's mass transit system is remarkably efficient and inexpensive. Approximately 40 percent of the country's paper and cardboard waste flow is now collected for recycling, and the government has set a target of recycling 80 percent by the year 2000. In order to achieve this, it has established such programs as one to reduce junk mail and one to increase paperless communication through computerization.

Singapore has also developed excellent information management systems, one of the key elements of a successful green plan. Like the Netherlands, it uses these systems to identify sources of pollution and other environmental problems.

Singapore's environmental regulations set tough standards and are strictly enforced. For example, heads of corporations are held responsible for environmental violations. If a problem occurs, a letter of warning is sent to the executive; if the problem is not corrected, the second letter of warning carries with it an automatic jail sentence for that executive. To date the government has issued about twenty first-warning letters, but has never had to send a second one.

Singapore has set some impressive long-term goals for itself, and I believe it will achieve them. A small, tightly knit nation, it may well be managing the problems of population density better than any other area in the world.

● ● ●

Two years after writing this book, I find myself just as excited by the potential of green plans. These immense efforts to recover environmental quality are succeeding in cities, states, and nations worldwide. They continue to convince me that green plans are the working definition of sustainability. The pioneers of green planning should be proud of their international leadership in an area of such critical importance. Solving the problem of environmental decline is a great challenge; now we have some examples to show us the way.

Notes

1. A Commitment to Change

1. Dutch Ministry of Housing, *The General Environmental Information Campaign*, p. 1. For more information on the Dutch public information campaign, see chapter 9.

2. All foreign currencies have been converted into U.S. dollars using the exchange rate current as of 21 October 1994. The Canadian green plan initially pledged $2.2 billion to be spent on the environment over a period of five years; faced with increasing budget deficits, the government later changed this to six years.

3. Johannes van Zijst, Counselor for Health and Environment, Royal Netherlands Embassy, personal conversation.

4. John Dewitt of the Save the Redwoods League, personal conversation.

2. Sustainability from Theory to Practice

1. World Commission on Environment and Development, *Our Common Future*, p. 43.

2. See Meadows, *The Limits to Growth*.

3. Ehrlich and Ehrlich, *Population, Resources, Environment*.

4. Daly and Cobb Jr., *For the Common Good*, p. 71.

5. Lester R. Brown et al, *State of the World 1994* (New York: W. W. Norton, 1994), p. 5.

6. Gow, "National Policy and Sustainable Development," p. 6.

7. Information in this section about New Zealand's RMA and the concept of sustainable management is from conversations with Lindsay J. A. Gow, Deputy Secretary for the Environment.

8. See Adriaanse, *Environmental Policy Performance Indicators*.

9. See Kroes, *Essential Environmental Information*, pp. 5–20.

10. National Institute of Public Health, *National Environmental Outlook 2*, p. 38.

11. Information regarding Canada's concept of sustainable development is from *Canada's Green Plan for a Healthy Environment*; R. W. Slater, Senior Assistant Deputy Minister, Environment Canada; and Brian Emmett, Assistant Deputy Minister,

Corporate Policy, Environment Canada.

12. Worster, *The Wealth of Nature*, pp. 153–54.

13. Worster, *The Wealth of Nature*, p. 153.

14. Vonkeman, *Highlights from the Environment*, pp. 41–42.

3. A Green Plan Predecessor

1. Seton, *California Energy*, p. 1.

2. California Department of Resources, *Investing for Prosperity: An Update*, pp. 1, 16.

3. Seton, *California Energy*, pp. 31–32.

4. Jeff Romm, personal conversation.

5. California Department of Resources, *Investing for Prosperity: An Update*, pp. 1–2.

6. Seton, *California Energy*, p. 5.

7. Resource Renewal Institute, *Investing for Prosperity: Results 1977–1987*.

8. Seton, *California Salmon*, preface.

9. Seton, *California Salmon*, preface.

10. Sin, *Evaluation of the California Forest Improvement Program*, p. 13.

11. Resource Renewal Institute, *Investing for Prosperity: Results 1977–1987*.

4. The Netherlands

1. *World Population Profile: 1991*, U.S. Department of Commerce, Bureau of the Census (Dec. 1991), pp. 27, 30.

2. Van Zijst, personal conversation.

3. Information on Queen Beatrix' speech is from Johannes van Zijst, who also provided much of the background on the events leading up to the passage of the NEPP and on the main elements of the NEPP itself.

4. Ministry of Housing, *National Environmental Policy Plan Plus*, p. 93, table 5.1.

5. Cahn, "Where Green Is the Color," p. 10.

6. Ministry of Housing, *National Environmental Policy Plan Plus*, p. 46.

7. Ministry of Housing, *To Choose or to Lose*, p. 12.

8. Van Zijst, personal conversation.

9. Van Zijst, personal conversation.

10. See Ministry of Housing, *To Choose or to Lose*, p. 94, p. 96, and p. 97, respectively.

11. See National Institute of Public Health, *National Environmental Outlook 2*.

12. Ministry of Housing, *Environmental Programme 1992–1995*, pp. 9–11.

13. Ministry of Housing, *Towards a Sustainable Netherlands*, p. 19 and p. 13, respectively.

14. Speech by J. J. de Graeff, Director of the Environment and Spatial Planning Office of the Federation of Netherlands Industry and the Netherlands Christian Employers'

Association, to a meeting of California business executives, August 1994.

15. See, for example, Marlise Simons' article "Dutch Do the Unthinkable: Sea Is Let In," *New York Times* (7 March 1993).

16. National Institute of Public Health, *National Environmental Outlook 2*, p. 353.

17. Ministry of Housing, *Towards a Sustainable Netherlands*, pp. 18, 20.

18. National Institute of Public Health, *National Environmental Outlook 2*, p. 31

19. Ministry of Housing, *Towards a Sustainable Netherlands*, pp. 24–25.

20. See Ministry of Housing, *Environmental Programme 1992–1995*.

5. New Zealand Starts from Scratch

1. Information regarding New Zealand's natural history and environmental problems is taken from Cronin, *Ecological Principles for Resource Management*; Gow, "National Policy and Sustainable Development"; and from personal communications with Mr. Gow, who also provided much of the background on the RMA.

2. Cronin, *Ecological Principles for Resource Management*, pp. 54–55.

3. Gow, "National Policy and Sustainable Development," p. 3.

4. Gow, "National Policy and Sustainable Development," p. 2.

5. Craig Lawson, Manager, Resource Management Directorate, New Zealand's Ministry for the Environment, personal conversation.

6. Information on the workings of the RMA comes from: Gow, "National Policy and Sustainable Development" and "New Zealand's Resource Management Act," and from personal communications with Mr. Gow.

7. Information on New Zealand 2010 provided by Roger Blakeley, Secretary for the Environment, personal conversation.

6. Canada's Green Plan

1. This and subsequent information regarding Canada's natural history and environmental problems is taken from *Canada's Green Plan for a Healthy Environment*.

2. Information on the circumstances surrounding the adoption of Canada's green plan comes primarily from a speech by R. W. Slater, Assistant Deputy Minister with Environment Canada, to Resource Renewal Institute's 1992 International Green Plan Conference, and from conversations with Mr. Slater.

3. From a prefatory statement to *Canada's Green Plan for a Healthy Environment* by Robert R. de Cotret, Canada's Minister of the Environment at the time of the green plan's publication.

4. Initially, the $2.2 billion was to be expended over the five-year lifespan of the first green plan; later, as Canada struggled with an economic recession and a change in government, that period was stretched to six years.

5. *Canada's Green Plan for a Healthy Environment*, pp. 5–6.

6. *Canada's Green Plan for a Healthy Environment*, p. 9.

7. This section on green plan goals and actions is based primarily on information provided by R. W. Slater in his speech to RRI's 1992 conference and in personal communications. Additional information comes from *Canada's Green Plan for a Healthy Environment.*

8. *Canada's Green Plan for a Healthy Environment*, p. 6.

9. Doering, *Canadian Round Tables on the Environment and the Economy*, p.9.

10. *Canada's Green Plan for a Healthy Environment*, p. 19

11. *Canada's Green Plan for a Healthy Environment*, p. 59 and p. 47, respectively.

12. Jean Charest, former Minister for the Environment, in a speech to the Environmental Forum, 10 December 1991.

13. *Canada's Green Plan for a Healthy Environment*, p. 5.

14. Mr. Slater's speech to RRI's 1992 conference.

7. On the Green Plan Path

1. Most of the information in this chapter regarding Norway and its environmental policies is from a presentation by Paul Hofseth, special advisor in the Norwegian Ministry of the Environment and a member of the Norwegian delegation to the UN Commission on Sustainable Development, to RRI's 1992 conference.

2. Norwegian State Pollution Control Authority, *The Future Is Now*, p. 4.

3. See, for example, World Resources Institute, *World Resources 1992–93*, p. 149, and Weterings, *The Ecocapacity as a Challenge*, p. 9.

4. Information on Sweden and its environmental policies is from a presentation by Per Kageson, consultant, to RRI's 1992 conference.

5. Information on Denmark's environmental policies is from the Netherlands' Ministry of Housing, *Comparison of Environmental Policy Planning;* Danish Ministry of the Environment, *Strategic Environmental Planning; Clean Is Better*, a brochure produced by Denmark's National Agency of Environmental Protection; and from information provided by the Danish government to the March 1994 meeting of the International Network of Green Planners in the Hague, the Netherlands.

6. See Danish Ministry of the Environment, *Strategic Environmental Planning.*

7. Most of the information on Austria and its environmental policies is from a presentation by Heinz Schreiber, Director General, Austria's Ministry for Environment, Youth, and Family, to RRI's 1992 conference.

8. Information on the NUP is from Austria's Federal Ministry for Environment, *Austrian National Environmental Plan.*

9. Most of the information on the United Kingdom and its environmental policies is from a presentation by John Stoker, head of the UK's Environmental Protection Central Division of the Department of the Environment, to RRI's 1992 conference, and from

personal communications with Mr. Stoker. Additional information from *This Common Inheritance*.

10. Germany's Federal Ministry for the Environment, *Environmental Protection in Germany*, p. 8. Most of the information on Germany's environmental policies is derived from this publication.

11. Most of the information on Singapore's policies is from Singapore's Ministry of the Environment, *The Singapore Green Plan*.

12. Commission of the European Communities, *Towards Sustainability*, p. 47.

13. Commission of the European Communities, *Towards Sustainability*, p. 48.

14. Commission of the European Communities, *Towards Sustainability*, p. 37.

15. Commission of the European Communities, *Towards Sustainability*, p. 64.

16. Commission of the European Communities, *Towards Sustainability*, p. 55.

17. Commission of the European Communities, *Towards Sustainability*, p. 81.

18. Commission of the European Communities, *Towards Sustainability*, pp. 141–42.

8. Broadening the Scope of Resource Management

1. Ministry of Housing, *Highlights of the Dutch National Environmental Policy Plan*.

2. From Per Kageson's presentation to RRI's 1992 conference.

3. Information on Dow's water recycling project provided by Randy Fischback, environmental compliance water manager at the Dow plant in Pittsburg, CA.

4. Kroes, *Essential Environmental Information*, p. 69, fig. 4.3.5.

9. A New Relationship between Government and Business

1. Information in this section is based on conversations with Maarten M. de Hoog, policy analyst with the Netherlands' Directorate for Industry, Building Trade, Products and Consumer Affairs, Directorate-General for the Environment, Ministry for Housing, Physical Planning and Environment, and with Johannes van Zijst.

2. Information regarding the covenant with the primary metals industry is derived from the agreement itself, "Declaration of Intent on the Implementation of Environmental Policy for the Primary Metals Industry," and from conversations with Mr. de Hoog.

3. "Declaration of Intent on the Implementation of Environmental Policy for the Primary Metals Industry."

11. A Greenprint for the United States

1. World Resources Institute, *World Resources 1992–93*, p. 346, table 24.1.

2. Information from a presentation by Thomas English of the Santa Clara County Manufacturing Group to a visiting delegation of Dutch officials, June 1993.

3. From a presentation by Margaret Dancey of United Technologies, June 1993.

4. See Idaho Department of Health and Welfare, *Mid-Snake River.*

5. Facts on File, Inc., *The New Book of World Rankings*, pp. 242–44.

References

General Books, Documents, and Articles

Cahn, Robert. "Where Green is the Color." *The Amicus Journal* (Fall 1992): 8–11.

Daly, Herman E. and John B. Cobb, Jr. *For the Common Good: Redirecting the Economy toward Community, the Environment, and a Sustainable Future.* Boston: Beacon Press, 1989.

Daly, Herman E. *Steady-State Economics.* San Francisco: W. H. Freeman and Company, 1977.

Ehrlich, Paul R., and Anne H. Ehrlich. *Population, Resources, Environment.* San Francisco: W. H. Freeman and Company, 1970.

Facts on File, Inc. *The New Book of World Rankings, Third Edition.* New York, 1991.

Goodland, Robert, et al., eds. *Environmentally Sustainable Economic Development: Building on Brundtland.* Paris: UNESCO, 1991.

MacNeill, Jim, Pieter Winsemius, and Taizo Yakushiji. *Beyond Interdependence: The Meshing of the World's Economy and the Earth's Ecology.* Oxford: Oxford University Press, 1991.

Meadows, Donella H., et al. *The Limits to Growth: A Report for the Club of Rome.* New York: Universe Books, 1972.

Nilsen, Richard. "Waiting for a U.S. Green Plan." *Whole Earth Review* (Summer 1992): 60-61.

Schmidheiny, Stephan, and the Business Council for Sustainable Development. *Changing Course: A Global Business Perspective on Development and the Environment.* Cambridge: MIT Press, 1992.

United Nations Commission on Environment and Development. *Agenda 21.* Boulder, CO: Earth Press, 1993.

World Commission on Environment and Development. *Our Common Future.* Oxford: Oxford University Press, 1987.

World Resources Institute, the United Nations Environment Programme, and the

United Nations Development Programme. *World Resources 1992–93.* New York: Oxford University Press, 1992.

_____. *World Resources 1994–95.* New York: Oxford University Press, 1994.

Worster, Donald. *The Wealth of Nature: Environmental History and the Ecological Imagination.* New York: Oxford University Press, 1993.

The Netherlands

Adriaanse, Albert. *Environmental Policy Performance Indicators.* The Hague: Ministry of Housing, Physical Planning, and Environment, 1993.

Kroes, H. W., ed. *Essential Environmental Information: The Netherlands 1991.* The Hague: Ministry of Housing, Physical Planning, and Environment, 1991.

Langeweg, Ir. F., ed. *Concern For Tomorrow.* Bilthoven: National Institute of Public Health and Environmental Protection, 1989.

Ministry of Housing, Physical Planning, and Environment. *Environmental News from the Netherlands* (quarterly newsletter).

_____. *Environmental Programme 1992–1995.* The Hague, 1991.

_____. *The General Environmental Information Campaign: Background and Main Features,* a working paper of the Central Public Information Division, 1991.

_____. *Highlights of the Dutch National Environmental Policy Plan,* brochure.

_____. *National Environmental Policy Plan Plus, 1990–1994.* The Hague, 1990.

_____. *National Environmental Policy Plan 2: Summary.* The Hague, 1994.

_____. *To Choose or to Lose: National Environmental Policy Plan, 1990–1994.* The Hague: SDU Publishers, 1989.

_____. *Towards a Sustainable Netherlands: Environmental Policy Development and Implementation.* The Hague, 1994.

National Institute of Public Health and Environmental Protection. *National Environmental Outlook 2: 1990–2010.* Bilthoven, 1991.

Verbeek, Jan. "A Groundswell Campaign for the Environment." Information and External Relations Department.

Vlak, Gus. "The Greening of the Netherlands." *Stanford Business School Magazine* (Dec. 1991): 4–5.

Vonkeman, Gerrit, ed. *Highlights from the Environment: Ideas for the 21st Century.* The Hague: Dutch Committee for Long Term Environmental Policy, 1991.

Weterings, R.A.P.M., and J. B. Opschoor. *The Ecocapacity as a Challenge to Technological Development.* Rijswijk: Advisory Council for Research on Nature and Environment, 1992.

New Zealand

Brash, David N. *Resource Management Ideas No. 2: Sustainable Management and the*

Environmental Bottom Line. Wellington: Ministry for the Environment, 1992.

Cronin, Karen. *Ecological Principles for Resource Management*. Wellington: Ministry for the Environment, Biological & Physical Systems Directorate, 1988.

Gow, Lindsay J. A. "National Policy and Sustainable Development: Fact or Fantasy." A paper presented to the State University of New York at Binghamton by the Deputy Secretary for the Environment, 1992.

_____. "New Zealand's Resource Management Act: Key Provisions and Their Implications." Address to Australian Planning Ministers' Conference by the Deputy Secretary for the Environment, 1991.

Ministry for the Environment. *Directions for Change: A Discussion Paper*. Wellington, 1988.

_____. *Global Environmental Issues and Sustainability: Proceedings of a Seminar Held in Wellington*. Wellington, 1989.

_____. *Resource Management: Guide to the Act*. Wellington, 1991.

Office of the Parliamentary Commissioner for the Environment. *Report on New Zealand Environmental Management 1987–91*. 1991.

Canada

Doering, Ronald L. *Canadian Round Tables on the Environment and the Economy: Their History, Form and Function,* working paper number 14, National Round Table on the Environment and the Economy. Ottawa, 1993.

Environment Canada. *Canada's Green Plan and the Earth Summit*. Ottawa, 1992.

_____. *Canada's Green Plan for a Healthy Environment*. Ottawa, 1990.

_____. *Canada's Green Plan: The Second Year.* Ottawa, 1993.

_____. *Canada's Green Plan: Summary*. Ottawa, 1990.

_____. *Canada's National Report for the United Nations Conference on Environment and Development, Brazil, June 1992*. Ottawa, 1991.

_____. *Economic Instruments for Environmental Protection: Discussion Paper*. Ottawa, 1992.

_____. *The State of Canada's Environment*. Ottawa, 1991.

Second National Stakeholders Assembly. *Project de Societé: Progress Report toward a National Sustainable Development Strategy for Canada*. Ottawa, 1993.

Slater, Robert W. "Changing the Way We Govern: Sustainable Development in Canada." Paper presented to the Environment Conference of the United Kingdom's Centre for Business and Public Sector Ethics, 1993.

California and the United States

California Department of Resources. *Investing for Prosperity: An Update*. Sacramento, 1982.

Comp, T. Allen, ed. *Blueprint for the Environment*. Salt Lake City: Howe Brothers, 1989.

Council on Environmental Quality and the Department of State. *Global 2000 Report to the President: Entering the Twenty-First Century*. New York: Penguin Books, 1982.

Florida Department of Environmental Regulation Office of Planning and Research. *Strategic Assessment of Florida's Environment: SAFE*. 1992.

Greenberg, Phillip A. *Toward a U.S. Green Plan: Thinking about a U.S. Strategy for Sustainable Development*. San Francisco: Resource Renewal Institute, 1992.

Idaho Department of Health and Welfare, Division of Environmental Quality. *Mid-Snake River Nutrient Management Project May 1993 Update*.

Louisiana Department of Environmental Quality. *LEAP to 2000*. 1991.

Resource Renewal Institute. *Investing for Prosperity: Results 1977–1987*. San Francisco, 1988.

Seton, Joel R. *The California Energy Conservation and Renewable Energy Resource Development Program 1980–88*. San Francisco: Resource Renewal Institute, 1988.

_____. *The California Salmon and Steelhead Restoration Program 1979–1988*. San Francisco: Resource Renewal Institute, 1988.

Sin, Meng Srun. *An Evaluation of the Performance of the California Forest Improvement Program*. A report submitted to the California Department of Forestry by Professor Srun of Humboldt State University's Department of Forestry, 1986.

Washington Department of Ecology, Washington Environment 2010 project. *Toward 2010: An Environmental Action Agenda*. 1990.

Other

Austria's Federal Ministry for Environment, Youth and Family. *Austrian National Environmental Plan (NUP): Interim Report 1993: An abstract of papers submitted by the sectoral working groups*. Vienna, March 1994.

Commission of the European Communities. *Towards Sustainability: A European Community Programme of Policy and Action in Relation to the Environment and Sustainable Development*. Luxembourg: Office for Official Publications of the European Communities, 1993.

Danish Ministry of the Environment. *Strategic Environmental Planning in Denmark*. Copenhagen, 1994.

Germany's Federal Ministry for the Environment. *Environmental Protection in Germany: National Report of the Federal Republic of Germany for the United Nations Conference on Environment and Development in June 1992 in Brazil*. Bonn, 1992.

Moiz, Azra. *The Singapore Green Plan – Action Programmes*. Singapore: Singapore's Ministry of the Environment, Times Editions Pte Ltd., 1993.

The Netherlands Ministry of Housing, Physical Planning and Environment. *Comparison of Environmental Policy Planning in Industrial Countries in the Context of the National Environmental Policy Plan.* Environmental Resources Limited. The Hague, 1990.

Norwegian Ministry of Environment. *Report to the Storting No. 46: Environment and Development.* 1989.

Norwegian State Pollution Control Authority. *The Future Is Now! A Summary of the State Pollution Control Authority's Long-Term Planning 1990–93.* Oslo, 1990.

This Common Inheritance: Britain's Environmental Strategy. London: HMSO, 1990.

Singapore's Ministry of the Environment. *The Singapore Green Plan: Towards a Model Green City.* Singapore: SNP Publishers, 1992.

Index

203

In the *Our Sustainable Future* series

Volume 1
Ogallala: Water for a Dry Land
John Opie

Volume 2
Building Soils for Better Crops: Organic Matter Management
Fred Magdoff

Volume 3
Agricultural Research Alternatives
William Lockeretz and Molly D. Anderson

Volume 4
Crop Improvement for Sustainable Agriculture
Edited by M. Brett Callaway and Charles A. Francis

Volume 5
Future Harvest: Pesticide-Free Farming
Jim Bender

Volume 6
*A Conspiracy of Optimism: Management of the National Forests
since World War Two*
Paul W. Hirt

Volume 7
Green Plans: Greenprint for Sustainability
Huey D. Johnson

Volume 8
Making Nature, Shaping Culture: Plant Biodiversity in Global Context
Lawrence Busch, William B. Lacy, Jeffrey Burkhardt,
Douglas Hemken, Jubel Moraga-Rojel, Timothy Koponen,
and José de Souza Silva

Volume 9
Economic Thresholds for Integrated Pest Management
Edited by Leon G. Higley and Larry P. Pedigo

Volume 10
Ecology and Economics of the Great Plains
Daniel S. Licht